Particulate
and Gaseous
Contamination
in Datacom
Environments

This publication was prepared in cooperation with TC 9.9, Mission Critical Facilities, Technology Spaces and Electronic Equipment.

Any updates/errata to this publication will be posted on the ASHRAE Web site at www.ashrae.org/publicationupdates.

Particulate and Gaseous Contamination in Datacom Environments

American Society of Heating, Refrigerating
and Air-Conditioning Engineers, Inc.

ISBN 978-1-933742-60-1

©2009 American Society of Heating, Refrigerating
and Air-Conditioning Engineers, Inc.
1791 Tullie Circle, NE
Atlanta, GA 30329
www.ashrae.org

Cover photograph used with permission of Crispin R. Semmens.

Library of Congress Cataloging-in-Publication Data

Particulate and gaseous contamination in datacom environments.

   p. cm. -- (ASHRAE datacom series ; bk. 8)

  Includes bibliographical references and index.

  Summary: "Provides the reader with information they need to maintain a high level of IT equipment dependability and availability by identifying datacom equipment susceptibility to particulate and gaseous contamination and the operational impact, as well as strategies for prevention, control, contamination testing, and analysis"--Provided by publisher.

  ISBN 978-1-933742-60-1 (softcover)

  1. Clean rooms. 2. Data processing service centers--Protection. 3. Electronic data processing--Equipment and supplies--Protection. 4. Air filters. 5. Dust control. 6. Contamination (Technology) 7. Decontamination (from gases, chemicals, etc.) I. American Society of Heating, Refrigerating and Air-Conditioning Engineers.

  TH7694.P37 2009

  697.9'316--dc22

2009016962

ASHRAE STAFF

SPECIAL PUBLICATIONS

Mark Owen
Editor/Group Manager
of Handbook and Special Publications

Cindy Sheffield Michaels
Managing Editor

James Madison Walker
Associate Editor

Amelia Sanders
Assistant Editor

Elisabeth Parrish
Assistant Editor

Michshell Phillips
Editorial Coordinator

PUBLISHING SERVICES

David Soltis
Group Manager of Publishing Services
and Electronic Communications

Jayne Jackson
Publication Traffic Administrator

PUBLISHER

W. Stephen Comstock

Contents

Acknowledgments

The information in this guide was produced with the help and support of the following corporations:

ANCIS® Incorporated

Beth Israel Deaconess Medical Center

Data Clean Corporation

Dell Computers

DLB Associates Consulting Engineers

Hewlett Packard

IBM

Intel Corporation

Lawrence Berkeley National Laboratory

Liebert Corporation

McKenney's Inc.

Microsoft

Rittal

Shen Milsom Wilke

Teradata

Uptime Institute

Verizon Wireless

ASHRAE TC 9.9 wishes to particularly thank the following people:

- **Rich Hill, Pam Lembke, David Moore, Arman Shehabi,** and **Prabjit Singh** for their participation as chapter leads, which included numerous conference calls, writing, and review.
- **Dr. Roger Schmidt** of IBM for his vision and thorough review of multiple drafts of this book.
- **Mr. Christian Belady** of Microsoft for his vision and leadership in the creation of this book.
- **Mr. John Quick** of IBM for his invaluable contributions in researching and writing material for this book.
- **Mr. Joe Prisco** of IBM for his overall leadership and editing of multiple drafts of this book.

In addition, ASHRAE TC 9.9 would like to thank the following people for their substantial contributions to the creation of this book: Tom Davidson, Bob Doherty, Ashok Gadgil, Srirupa Ganguly, Magnus Herrlin, Greg Jeffers, Bob McFarlane, Mike Patterson, Dave Quirk, Fred Stack, Bob Sullivan, Bill Tschudi, Herb Villa, and David Wang.

1

Introduction

Datacom equipment center owners and operators focus much of their attention on the physical structure and performance of the datacom infrastructure environment (e.g., power, cooling, and raised-access floor equipment). However, today's intricate and sensitive information technology (IT) equipment (also called *datacom equipment* or *computer equipment* throughout the book) requires a certain level of environmental control for gaseous and particulate contamination that is present within the facility's datacom equipment center environment. Datacom equipment center contamination is frequently overlooked and, if left unrestrained, can degrade the reliability and the continuous operation of mission-critical IT equipment within a facility.

To maintain a high level of IT equipment dependability and availability, it is critical to view contamination in a holistic way. It should be acknowledged that the datacom equipment center is a dynamic environment where many maintenance operations, infrastructure upgrades, and IT equipment change activities occur on a regular basis. Airborne contaminants harmful to sensitive electronic devices can be introduced into the operating environment in many ways during these and other activities. The fundamental focus areas that necessitate examination start with the outdoor ambient air pollutants surrounding the facility. Outdoor air that purposely enters the building for datacom equipment center free cooling, datacom equipment center positive pressurization, or human occupancy air changes must be filtered and possibly conditioned. Once inside the building, maintenance operations within the building's environmental envelope and the datacom infrastructure equipment itself must be considered. Datacom workers also add contamination from hair, lint on clothing, and other contaminants tracked in on footwear to the datacom equipment center. With proper planning and controls, datacom equipment center operators can minimize contamination and potential negative effects in the datacom equipment center.

Datacom managers and operators should include a datacom equipment center environmental contamination section as part of the standard operating procedure.

The association between contamination and hardware failures is often overlooked. Occasionally, the absence of contamination controls results from cost-cutting actions or from lack of knowledge. Particle and gaseous contamination can result in intermittent equipment glitches or in unplanned shutdowns of critical systems that often mean significant business and financial losses. Examples of contamination events are provided throughout the book. In many cases, the events are written generically to illustrate points that support the text.

The intent of this publication is to provide basic information that is essential to the control and prevention of particulate and gaseous contamination within datacom facilities. Understanding the critical parameters outlined in this publication will provide equipment manufacturers and facilities operations personnel with a common set of guidelines for contamination control that can enhance the longevity of datacom equipment. The book does not cover issues related to contamination and filtration of open water systems, such as condenser water systems, used in datacom environments.

The intended audience for this publication is:

- planners and datacom facility operation managers
- datacom facility architects and engineers who require insight on datacom environmental controls for gaseous, organic, and particulate contamination
- datacom facility service providers
- datacom equipment manufacturers

1.1 GENERAL DESCRIPTION OF PARTICULATE MATTER

Particulate matter (PM) refers to small solid or liquid particles that can become airborne with different airborne lifetimes. For the purposes of this book, the terms *particle, particulate, aerosol,* and *dust* will be considered equivalent and all are represented by the term *PM*. The size of PM spans a vast size range from 0.001 to more than 100 micrometers (μm) (3.93701×10^{-8} to more than 0.003937008 in.) in diameter. The United States Environmental Protection Agency (EPA), which monitors PM from a health point of view, categorizes particle mass concentration as PM2.5 and PM10, representing particles smaller than 2.5 and 10 μm (9.84252×10^{-5} and 0.000393701 in.), respectively. More specifically, PM can be categorized in three size modes: fine mode (0.001–0.1 μm [3.93701×10^{-8}–3.93701×10^{-6} in.]), accumulation mode (0.1–2.5 μm [3.93701×10^{-6} –9.84252×10^{-5} in.]), and coarse mode (2.5–10 μm [9.84252×10^{-5}– 0.000393701 in.]), which is often limited to particles smaller than 10 μm (0.000393701 in.) but can include fibers and particles as large as 100 μm (0.003937008 in.). PM in each of these size categories may be composed of different materials, come from different sources, and vary in airborne suspension lifetime. See Figure 1.1 for typical size ranges of PM (Bell 2000). Consequently, PM

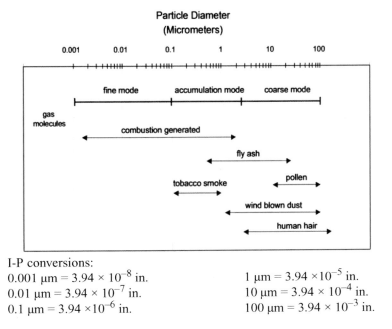

I-P conversions:

$0.001 \ \mu m = 3.94 \times 10^{-8}$ in. $1 \ \mu m = 3.94 \times 10^{-5}$ in.

$0.01 \ \mu m = 3.94 \times 10^{-7}$ in. $10 \ \mu m = 3.94 \times 10^{-4}$ in.

$0.1 \ \mu m = 3.94 \times 10^{-6}$ in. $100 \ \mu m = 3.94 \times 10^{-3}$ in.

Figure 1.1   Size ranges of various PM sources.

in each size category is associated with different equipment reliability concerns. This book discusses particles in each of these size modes.

1.2   GENERAL DESCRIPTION OF GASEOUS CONTAMINATION

Gaseous contamination is an impurity in the air that has an adverse effect on computer hardware. The gases can occur naturally or can be by-products of industrial or manufacturing processes. Gases tend to occupy an entire air volume uniformly at standard room pressure and temperature. The gases can either act alone or in conjunction with other gases or PM to form compounds that can result in oxidation on metallic materials. The oxidation results from a chemical reaction, which causes irreversible destruction on the surface of a circuit board, on the leads of a connector, or on pins of an integrated circuit.

Gaseous contamination in the form of outside pollution (e.g., smog), particularly when composed of sulfur, bromine, and chlorine compounds, can corrode power and cooling equipment, electronic circuit boards, and datacom connections, which over time will degrade the overall reliability of datacom equipment.

1.3 CONTAMINANT SOURCES

PM is generated both naturally and by humans (anthropogenically). In the outdoors, fine-mode (0.001–0.1 µm [3.93701×10^{-8}–3.93701×10^{-6} in.]) and accumulation-mode particles (0.1–2.5 µm [3.93701×10^{-6}–9.84252×10^{-5} in.]) are formed primarily during combustion processes. Significant urban PM sources include automobiles, electricity generation, and wood-burning fireplaces (Seinfeld and Pandis 1998). Larger particles can include sea salt, natural and artificial fibers, plant pollens, and wind-blown dust. Accumulation-mode particles are the PM of principal concern when outdoor air is entering the datacom equipment center environment. Conventional filters in mechanical ventilation systems are able to very efficiently remove PM in the coarse mode (> 2.5 µm [> 9.84252×10^{-5} in.]). Figure 1.2 shows the minimum efficiency reporting value (MERV) filter efficiency of MERV 9 and MERV 13 filters as a function of particle size (Hanley et al. 1994; Riley et al. 2002). More information about air filtration can be found in Chapter 4. Particles with diameters larger than about 1.0 µm (3.93701×10^{-5} in.) are effectively removed by filtration since the mass of these particles prevents them from following the quickly changing airflow pattern of air moving across the filter. This causes these larger particles to crash into the filter fibers. Particles smaller than about 0.1 µm (3.93701×10^{-6} in.) are also effectively removed by filtration since the small size of these particles causes them to drift away from the air maneuvering through the filter and collide with the filter fibers. Particles in the in-between range (the accumulation-mode size range) are most difficult to remove, because these particles most easily follow the airflow through the filter fibers. Outdoor particles beyond the accumulation-mode size range can enter datacom equipment centers through openings in the building envelope, but datacom equipment centers can be positively pressurized to avoid this form of particle infiltration. However, one study indicates that introduction of outdoor air for the sole purpose of guarding against infiltration actually carries more particles into the building than the added pressure keeps out (Herrlin 1997).

Previous measurements have shown outdoor air to be the main source of PM in the data center environment (Shehabi et al. 2008), but anecdotal evidence suggests that episodic indoor events can potentially contribute to indoor PM concentrations. Indoor-generated PM can include particles in all size ranges. Fan belt wear in air-handling units (AHU), toner dust from copiers, and printers can be sources of coarse-mode particles. Coarse-mode particles can also come from occupant hair and clothing or occupant activities, such as the unpacking of equipment or construction (e.g., cement dust, drywall dust, insulation, paper/cardboard fiber). Small fibers from occupants or occupant activities represent particles greater than 10 µm (0.000393701 in.) that can be of concern. Other documented fibers include small zinc formations, commonly referred to as *zinc whiskers*, which have been observed forming on the surface of zinc electroplated steel.

Gases entering the datacom equipment center from the outdoors can react once inside the environment and form particles in the fine- and accumulation-mode size range.

1.4   HOW CONTAMINANTS SETTLE ON EQUIPMENT

PM is transported with the movement of air, also known as *advection*, throughout the datacom equipment center. Computer room air-conditioning (CRAC) units or computer room air-handling (CRAH) units often provide air conditioning in datacom equipment centers, where either the CRAC or CRAH is situated on the data center floor. Datacom equipment centers can be raised-access floor or non-raised-access floor environments. Figure 1.3 shows the airflow in a typical raised-access floor data center. Outside air, also referred to as *makeup air*, enters the side of the AHU, where it passes through a series of filters and is conditioned. The conditioned air is then supplied to the servers through the raised floor. Fans within the datacom equipment pull air through the

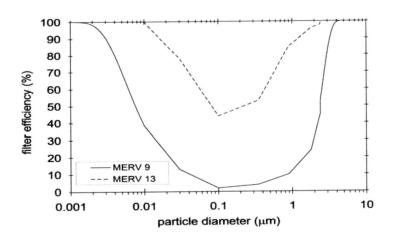

I-P conversions:
0.001μm = 3.94 × 10–8 in.   1 μm = 3.94 ×10–5 in.
0.01 μm = 3.94 × 10–7 in.   10 μm = 3.94 × 10–4 in.
0.1 μm = 3.94 ×10–6 in.   100 μm = 3.94 × 10–3 in.

Figure 1.2   Filter efficiency of commonly used filters in datacom environments.

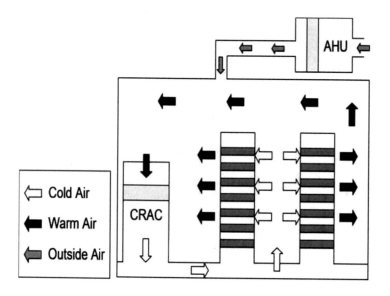

Figure 1.3 Airflow in a raised-access floor data center.

cabinets. The warmed air eventually makes its way back to the intake of the CRAC unit or the AHU return (not shown in Figure 1.3). Most air circulation in datacom centers is internal to the datacom center zone. The majority of datacom centers are designed to have only a small portion of outside air enter the datacom center for positive pressurization. Some datacom centers provide no ductwork for outside air to directly enter the datacom center area. Rather, outside air is only provided by infiltration from adjacent zones, such as office spaces or hallways. However, a growing number of datacom centers use air handlers designed with air-side economizers (similar to those used in commercial buildings) to take advantage of the energy- efficiency benefit of using a high volume of outside air for cooling. It should be noted that the traditional closed-flow layout results in the same air circulating through filters repeatedly. The intake of large volumes of external air and its subsequent exhaust from the datacom equipment center results in fewer passes through the filtration system for any given volume of air.

PM deviates from airflow paths and settles onto interior surfaces primarily through three different mechanisms:

- gravitational settling
- diffusional movement
- electrostatic attraction

Gravitational settling is a function of particle mass and has the greatest influence on large particles. Gravitational settling becomes insignificant for particles smaller than approximately 1 μm (3.93701×10^{-5} in.) in diameter, while particles greater than 10 μm (0.000393701 in.) have short airborne residence times due to the strong gravitation forces that cause particles to quickly settle onto horizontal surfaces. Strong air currents can prolong airborne residence times, even for large particulates.

The diffusional movement of PM is caused by random collision of air molecules against airborne particles. The result of these collisions allows particles to migrate from higher particle concentrations to lower particle concentrations. Diffusion is significant only with very small particles and has minimal influence on particles greater than 0.1 μm (3.93701×10^{-6} in.) in diameter. Diffusion affects particles equally in all directions. While larger particles primarily deposit onto horizontal surfaces, smaller particles have an equal tendency to deposit on either horizontal or vertical surfaces.

Electrostatic attraction is the force between opposite charges that pulls particles together or causes them to settle on surfaces. The simplest example is static electricity, the same force which may cause clothes to cling together.

Once a particle comes into contact with a surface, either by gravitational settling, diffusional movement, or electrostatic attraction, it is generally expected to remain deposited. Resuspension of particles in the air is expected to be minimal because of the cohesive forces between the particles and surface. Mechanical processes such as floor sweeping, movement of floor or ceiling tiles, or equipment maintenance often cause resuspension.

Gaseous contaminants diffuse in the air to occupy the entire volume, but differences in air density may cause stratification. The movement of the contaminants is relative to the air movement.

1.5   DIFFERENCES BETWEEN HUMAN HEALTH AND DATACOM EQUIPMENT CONCERNS

PM vulnerabilities in humans are very different than those in datacom equipment. Most studies of PM in the workplace are focused on the range of particulates harmful to humans. Two frequently tracked values are PM2.5 and PM10. These are the airborne particle sizes of PM < 2.5 μm (PM < 9.84252×10^{-5} in.) and PM < 10 μm (PM < 0.000393701 in.) in diameter. These values roughly relate to material sizes remaining after various stages of the human filter. Human physiology provides multiple filtration stages as PM-laden air is drawn into the respiratory system. Filtration in the nose and throat removes most large airborne particulates.

Many electronic assemblies are vulnerable to much larger airborne material. Long fibers can be particularly problematic for electronic equipment. Fibers in excess of 5 mm (0.196850394 in.) in length have been found within IT equipment.

1.6 OVERVIEW OF CHAPTERS

Chapter 1—The introduction states the purpose/objective of the publication as well as some background information about the importance, sources, and nature of PM and gaseous contamination. This chapter also includes a brief overview of the remaining chapters and their relationships to each other.

Chapter 2—This chapter discusses PM and gaseous contaminants and how their interaction can impact the reliable operation of datacom equipment. The chapter examines the datacom equipment components, devices, and subsystems vulnerable to PM and gaseous contaminants and describes their mechanical, chemical, and electrical effects. Some of the more pronounced effects are overheating, corrosion, and arcing. These effects can have serious consequences on the long-term health of the datacom equipment.

Chapter 3—There are several test methods, guidelines, and limits available for consideration and use in datacom environments. This chapter examines pertinent documents but does not recommend a single set of limits. Because of the complexity of PM and gaseous impacts on datacom equipment, individual datacom manufacturers may choose to either adopt a single industry work, adopt a combination of industry works, or establish their own set of design requirements. Establishing individual test methods and limits is done to address specific materials, electrical design characteristics, and/or thermal design characteristics necessary to meet performance or other critical design objectives.

Chapter 4—Best practices for facility level prevention and controls are documented in detail in this chapter. PM and gaseous exposure risks and hazards need to be identified so that datacom equipment centers can be designed and constructed to keep contamination out. Contaminant prevention is a very important consideration with respect to facility location, design, construction, and the cooling system. Even with contamination mitigation through prevention, potential PM and gaseous contamination sources are unavoidable and must be controlled. Actionable recommendations, aimed at controlling PM and gaseous contamination, are provided in this chapter to be incorporated into general business processes and procedures. Implementing facility prevention and control techniques can diminish PM and gaseous contamination to levels below those that most manufacturers would be concerned about in a typical datacom equipment center. One important aspect of control is monitoring to evaluate the condition of the datacom equipment center. Some monitoring is as simple as visual inspection, but at times, testing and analysis is necessary. Examples are documented in Chapter 5.

Chapter 5—Testing and analysis can be used to conclusively determine the presence and quantity of PM and gaseous contaminants. This chapter presents methodical, comprehensive surveys that can be used to determine the presence of contaminants. Some of the site surveys can be done visually, but typically instruments are required to collect data and professional analysis is necessary to interpret the results and identify sources for remediation.

Chapter 6—Air-side economizers represent a significant opportunity for datacom equipment centers to save energy. This chapter offers a set of considerations for implementing air-side economizers successfully. It also identifies potential threats to datacom equipment centers.

Chapter 7—This chapter offers a short summary of key ideas presented in this book.

Appendix A—This appendix discusses proposed datacom environment contamination levels.

Appendix B—This appendix examines field contamination occurrences.

Appendix C—This appendix presents future work in PM and gaseous contaminants.

2

Datacom Equipment Vulnerability

2.1 INTRODUCTION

Airborne contamination is a fact of life in most datacom equipment center environments. In a well-provisioned datacom equipment center with properly maintained filtration, the concentrations of airborne contaminants should pose little or no danger to the datacom and infrastructure equipment. However, not all usage environments are identical. When considering typical datacom equipment center environments, there is reason for equipment vulnerability concerns. Even properly filtered datacom equipment center environments contain some level of particulate matter (PM) in the air. Gaseous contaminants also have the potential to cause difficulties for datacom equipment. The potential for adverse effects is increased when gaseous contaminants and particulate contaminants combine and high humidity levels occur. A minute amount of PM or gaseous contamination can introduce intermittent malfunctions, undesirable changes in electrical characteristics, or complete equipment failures.

Some regions of the world pose a greater risk of particulate contamination. Differences arise from outdoor contamination levels, facility design practices, and operational methods. There is also a trend toward deployment of datacom equipment center infrastructure and datacom equipment in less favorable environmental conditions. Appendix A describes a proposed datacom equipment center classification standard that highlights different exposure risks to PM and gaseous contamination for five datacom equipment center facility types.

This chapter examines how datacom equipment center equipment is vulnerable to airborne particulates and gaseous contaminants. This discussion emphasizes the need for monitoring contaminant levels and the need for a well-maintained datacom equipment center environment.

2.2   REASONS FOR INCREASED CONCERN

Trends toward compact datacom equipment affect many factors that make airborne contamination a more significant issue than it has been in the past. Higher power densities within air-cooled equipment require extremely efficient heat sinks and greater air exchange volume increasing the airborne contaminant exposure. Some newer high-density cooling systems are highly efficient when clean but may readily clog with airborne PM, resulting in degradation of both cooling and power efficiency. PM can also degrade high-density electronic interconnects that require tight control of electrical impedance. Dense electronic component terminal spacing also increases the probability for conductive bridging due to PM or dendrite growth.

2.2.1   Restriction of Hazardous Substances

Directive 2002/95/EC of the European Parliament and of the Council of 27 January 2003 on the Restriction of the use of Certain Hazardous Substances on Electrical and Electronic Equipment (European Parliament 2003) eliminated lead from electronic equipment, requiring manufacturers to find new manufacturing processes. Some of the replacement processes may be more vulnerable to gaseous contamination. For example, creep corrosion has been experienced in environments where lead-based processes performed satisfactorily. As of late, manufacturers of electronic equipment have characterized these failure mechanisms and are incorporating design rules, test methods, and acceptable surface finishes to improve equipment reliability (Schueller 2007).

2.3   AIRBORNE CONTAMINATION INGRESS MECHANISMS

PM and gaseous contamination can enter datacom equipment in several ways. In most datacom equipment, there are two significant ingress mechanisms for PM. Suspended PM can be drawn in with inlet cooling air that circulates throughout the equipment or PM can settle on equipment surfaces by sedimentation. Both contamination mechanisms require the PM to be airborne, which provides a practical limit to particle size and mass. Air velocity and turbulence also affect the amount of and time aloft for suspended material.

Gaseous contamination enters equipment by convection or dispersion through any datacom or infrastructure equipment opening (e.g., sheet metal seams). The composition and concentration of gases, materials present, moisture content, temperature, and other factors in combination must be considered to understand the potential effects.

2.4   PARTICULATE MATTER PROPERTIES AND EFFECTS

PM can be composed of organic, inorganic, synthetic, or metallic materials, alone or in combination. PM can take various shapes: granular, fibrous, plate-shaped, or irregular. PM can have abrasive, corrosive, electrically or thermally conductive, electrically or thermally insulative, or hygroscopic properties. PM can interfere with airflow and optical signaling. Accumulation of PM is also cosmetically undesirable.

The effects of PM contamination on datacom equipment can generally be broken into the following categories:

- mechanical effects—include obstruction of cooling airflow, interference with moving parts, abrasion, optical interference, interconnect interference, or deformation of surfaces (e.g., magnetic media) and other similar effects
- chemical effects—include corrosion of surfaces, dendrite growth, and material property changes such as embrittlement or optical clouding of surfaces
- electrical effects—include impedance changes and electronic circuit conductor bridging

These PM contamination effects may occur alone or in combination with one another. Some effects may take place from only one form of airborne contamination; others may require combinations of materials.

2.4.1 Areas Susceptible to Particulate Matter Accumulation

Datacom electronic and infrastructure equipment can be degraded in many ways by PM. Practically all of the properties listed in the previous section can cause operational difficulty for electronic equipment.

In general, the amount of PM accumulation is directly related to the volume of air exchanged through the equipment. For this reason, forced-air-cooled equipment may be more susceptible to PM accumulation than liquid-cooled or free-convection-cooled equipment. Not all particulates entering equipment will be captured by the equipment; some particles will pass through and exit with the exhaust air.

Some areas within air-cooled equipment are more likely to collect PM than others. The following are examples of typical accumulation locations:

- small airflow openings, such as intake and exhaust vents, including unintended air leakage areas like rivet holes and sheet metal seams
- fine-pitch heat sink cooling features
- areas where flow bypass is not possible (particularly noticeable where flow ducting is used to force air through a heat sink)
- areas where airflow undergoes a sudden reduction in speed or change in direction
- sharp, rough, or adhesive surfaces, including surfaces made rough and/or adhesive by airborne contaminants

2.4.1.1 Air Intake and Exhaust

PM accumulation can restrict air intake and exhaust ventilation openings. Restriction of these openings may increase pressure drop across ventilation openings and result in reduced airflow through the system or increased fan load.

2.4.1.2 Fans

Fans themselves are not particularly vulnerable to PM. However, PM often affects fans indirectly. Fans installed directly on heat sinks may be affected by PM accumulation on the heat sink, which can eventually block fan rotation. However, PM often affects fans indirectly. Restrictions of intake and outlet vents can alter a datacom equipment pressure curve, resulting in greater fan loading. Changes in processor temperature as a result of blocked heat sinks can also trigger fan speed increases in systems equipped with fan speed controllers. These contamination issues may be misdiagnosed as fan problems.

2.4.1.3 Heat Sinks and Cooling Mechanisms

One significant issue for air-cooled equipment is the effect of PM on cooling efficiency and airflow through heat sinks. Existing microprocessor heat sink designs often use dense stacks of thin metal plates. The narrow airflow channels in these heat sinks are especially susceptible to blockage by airborne PM. PM with significant fiber content is especially challenging. Fibers can become trapped against the leading edges of heat sinks, bridging the gaps between the fin plates. When additional fibers contact the initial fibers, they often become entwined. This process eventually results in a network of tangled fibers that acts as a trap for smaller particulates. This process results in an unintended filter that rapidly grows until a nearly complete airflow blockage occurs through the heat sink passages. This process significantly degrades the cooling efficiency of the heat sink.

2.4.1.4 Magnetic Media and Optical Drive Mechanisms

Magnetic media drives are vulnerable to a number of contamination-related failure mechanisms. PM can accumulate in grease used to lubricate positioning and auto-loading mechanisms. This can cause abrasion or blockage of the moving parts. Some magnetic media can create its own contamination through oxide flake off. Particles can cause deformation of the media surface or interfere with head contact or spacing. In extreme cases, PM may abrade read/write heads. Fixed-disk (hard) drives typically employ filters to avoid these vulnerabilities. Optical drives may experience similar abrasion or blockage of moving parts. In addition, PM may interfere with the optical signal used to read and write data on the optical media.

2.4.1.5 Electrical Signals and Interconnects

PM with electrically conductive properties may provide unintended electrical paths within the datacom equipment. In extreme cases, airborne PM such as zinc whiskers or other conductive materials may form low-impedance short circuits within the datacom equipment. Other effects may be less obvious, such as leakage paths provided by moisture-laden hygroscopic particulate accumulation. Of equal concern are electrically insulative particles, which can interfere with connector contacts. Insulative particles may cause increased contact resistance or even open circuits. These effects may be intermittent in nature and can be very hard to diagnose.

PM composed of water-soluble ionic salts poses a concern in datacom equipment centers. If these salts absorb sufficient moisture from the ambient air, they can produce electrically conductive contaminants. Sulfate, nitrate, and sea salt (NaCl) particles are the most common water-soluble ionic salts in ambient air and represent a significant portion of urban PM (McMurry et al. 2004). Empirical results have shown that exposure to high sulfate concentrations at high humidity can cause electronic equipment failure (Litvak et al. 2000). While particle accumulation may occur on a time scale of years, the change in electrical conductivity can happen abruptly. Sudden spikes in relative humidity have the potential to induce equipment failure.

Figures 2.1 through 2.7 illustrate examples of particulate accumulation within datacom equipment. These examples are from accelerated testing under extreme

Figure 2.1   PM accumulation in a fine-pitch heat sink. Fibers have bridged gaps between fins and are now trapping additional material. This heat sink will continue accumulating PM until it is completely covered. The deposited material becomes an increasingly efficient filter. Cooling efficiency is reduced as air passages are blocked.

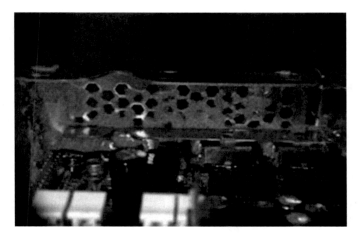

Figure 2.2 Exit vents viewed from inside a system (cover removed). PM accumulation blocks a significant portion of the available area. Notice there is not much accumulation of the printed circuit board itself.

Figure 2.3 An example of accumulation around unexpected intakes. PM is shown in the small gap surrounding the connector sockets.

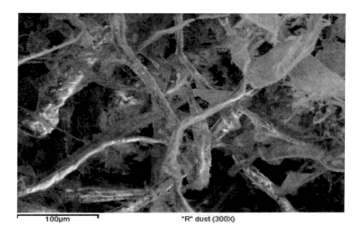

100µm   "R" dust (300X)

Figure 2.4 Scanning electron microscope (SEM) image of PM accumulation from a heat sink. Also, notice how the intertwined fibers form a matrix that traps smaller particulates. Notice that most of the material is far larger than 10 microns (3.94×10^{-4} in.).

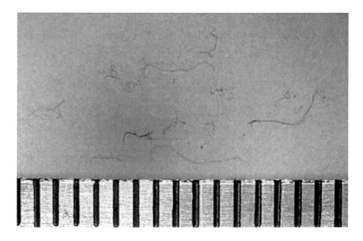

Figure 2.5 These fibers were recovered from a heat sink after exposure to field use conditions. The scale markings at the bottom of the photograph are 1 mm (0.04 in.) each. Some fibers are 5 mm (0.20 in.) in length.

Figure 2.6 This optical photograph shows the copper sulfide bridging between the integrated circuit leads and a contact on the card.

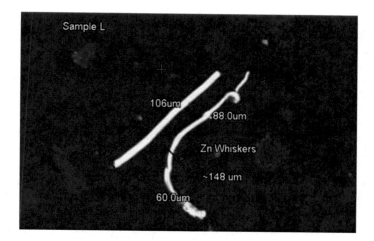

Figure 2.7 A conductive substance frequently found on the underside of wood-core floor panels with flat, zinc-coated steel bottoms. Zinc whiskers are typically several micrometers in length. Airborne introduction of zinc whiskers into datacom equipment may result in short circuits.

conditions, but similar patterns have been observed in contaminated datacom equipment center installations.

2.5 GASEOUS CONTAMINATION

The effects of gaseous contamination, alone or in combination with PM, may cause significant damage to datacom equipment.

2.5.1 Gas Properties

The following information, reproduced from *Design Considerations for Datacom Equipment Centers*, Second Edition (ASHRAE 2009a), provides the typical characteristics of gases that result in deterioration of metals in computer hardware.

Corrosion in an indoor environment is most often initiated by a short list of compounds or combinations of a few compounds. However, there are literally hundreds of compounds that can have a corrosive effect that are not normally found in an indoor environment. Table 2.1 includes the most common and abundant corrosion-inducing compounds that might be found in an indoor datacom environment. The compounds present are highly dependant on the controls put in place to mitigate them.

2.5.1.1 Corrosion Risks From Airborne Contamination

Sulfur-bearing gases, such as sulfur dioxide (SO_2) and hydrogen sulfide (H_2S), are the most common gases in datacom centers that cause hardware corrosion. An example of a component failure as the result of sulfur-bearing gases emanating from volcanic activity is shown in Figure 2.8. The device in Figure 2.8 failed because sulfur-bearing gases corroded the silver metallization. The bottom half of the figure shows silver sulfide flowers extruding from under the dielectric insulation. The sulfur-bearing gases entered the component package and attacked the silver resulting in the formation of silver sulfide (Ag_2S). The mechanical pressure created by the Ag_2S formation inside the package damaged the mechanical integrity of the package and caused the device to fail.

Solid PM accumulation can also cause corrosion in datacom equipment. Accumulated PM composed of chemically active substances, such as water-soluble ionic salts, may react chemically with datacom equipment components. Many PM compounds absorb moisture from the ambient air and may produce corrosive compounds when saturated. The amount of humidity required to cause saturation varies with the compounds involved.

In summary, airborne PM and gaseous contamination can cause a wide range of unwanted datacom equipment center conditions that can affect the continued operation and reliability of datacom equipment. Degradation will vary depending on

Table 2.1 Compounds of Most Concern
in the Datacom Equipment Center

Corrosive Material and Formula	Physical Characteristics	Typical Corrosion Initiators	Typical Industrial Sources
Sulfur dioxide (SO_2)	Colorless gas, irritating pungent odor	Reacts with water to produce highly corrosive sulfuric acid	Product of combustion of fossil fuel and incineration of organic waste; also found in paper, fabrics, food preservation, fumigants, and refining
Hydrogen sulfide (H_2S)	Colorless gas, odor of rotten eggs	Forms metallic sulfides	Intermediate in chemical synthesis
Chlorine (Cl)	Greenish-yellow gas, pungent, initiating, choking odor	Very reactive and produces corrosive metal salts, combines with all elements except carbon and noble gases	Widespread use in chemical synthesis, bleaching, oxidation
Hydrogen Chloride (HCl)	Colorless, corrosive gas, pungent characteristic odor, fumes in air	Quickly soluble in water reacting to form hydochloric acid, corrosion products are copper chloride and other metal salts	Chlorides and HCl are by-products of coal and incinerator combustion; widespread use in chemical synthesis, polymers, rubber, and pharmaceuticals
Nitrogen Dioxide (NO_2)	Reddish brown gas, suffocating odor	Highly reactive, forms acid with water, corrodes electronic materials forming highly corrosive nitric acid	Used in chemical synthesis and explosives, product of combustion from energy production and automobiles
Ozone (O_3)	Characteristic pleasant odor in small concentrations	Most reactive form of oxygen, found in smog	Disinfectant for water and air; bleach for textiles, waxes, and oils
Ammonia (NH_3)	Colorless, corrosive alkaline gas, pungent odor	Readily dissolves in water and combines readily with acid gases, producing a corrosive salt	Refrigeration, fertilizers, synthetic fibers and plastics; widely used in chemical synthesis

Source: *Design Considerations for Datacom Equipment Centers*, Second Edition (ASHRAE 2009a)

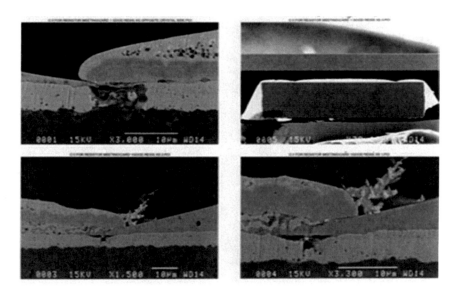

Figure 2.8 The top left and top right micrographs show the resistor in low magnification. The bottom left and bottom right micrographs show silver sulfide flowers protruding out from under the dielectric insulation. The resistor terminal was electrically undermined by the formation of silver sulfide.

equipment location and the chemistry, quantity, and composition of contaminants. Equipment issues arise from mechanical, chemical, or electrical degradation. Issues are often intermittent and difficult to diagnose. This is due, in part, to the interaction of multiple factors such as particulates, gases, humidity, and other environmental conditions that combine to trigger noticeable effects.

To a certain extent, airborne contamination in datacom environments is inevitable. Nonetheless, every effort should be used to minimize airborne contamination, including gaseous and particulate contaminants. Contamination control actions and recommendations are discussed in Chapters 5 and 6.

3

Industry Specifications and Guidelines

3.1 INTRODUCTION

This chapter summarizes typical cleanliness guidelines and limits intended for commercial and industrial environments to quantify tolerable exposure to particulate matter (PM) and gaseous contaminants. There are several published and accepted standards relevant to commercial and industrial environments that are extended to the datacom environment. Using multiple standards can make it difficult to establish a single set of contamination limits for the datacom industry, as contamination, concentration, composition, humidity, and thermal environment interactions vary within a facility. Also, one contaminant by itself may not be a significant problem, but when it is atomically combined with other environmental contaminants or with moisture, the combination may result in datacom equipment failures. Product design of datacom equipment is another related aspect since the materials, airflow paths, and types of cabinets (e.g., liquid-cooled or self-contained cabinets) can determine the susceptibility to PM and gaseous contaminants.

This chapter summarizes available industry works and their limits for PM within their scope. The telecom industry generally follows Network Equipment-Building System (NEBS) standards for network switching facilities. The IT industry may use International Organization for Standardization's (ISO) *ISO 14644-1, Cleanrooms and Associated Controlled Environments—Part I: Classification of Air Cleanliness* (ISO 1999) for computer rooms and data centers. Equipment manufacturers may design to International Electrotechnical Commission's (IEC) *IEC 60721-3-3, Classification of Groups of Environmental Parameters and Their Severities—Stationary Use at Weather-Protected Locations* (IEC 2002). In addition to these standards publications, more research by academia is necessary to conclusively determine the impact of PM and gaseous contaminants on IT equipment.

3.2 PUBLISHED GUIDELINES AND LIMITS FOR PARTICULATE MATTER

3.2.1 GR-63-CORE/Network Equipment-Building Systems— Telecommunications

GR-63-CORE, Network Equipment-Building Systems (NEBS) Requirements: Physical Protection (Telcordia 2006) is a set of physical requirements and testing standards developed for network-switching centers. The NEBS guidelines were originally developed by Bell Telephone Laboratories in the 1970s and were expanded by Bell Communications Research (Bellcore). Bellcore was later renamed Telcordia Technologies, Inc. The NEBS requirements were created to ensure that equipment operates under the range of temperature, humidity, vibration, and airborne contamination levels within network facilities. Telecommunications companies generally require their datacom equipment providers to comply with the NEBS de-facto standard. The NEBS requirements for airborne contamination levels provide specific concentration limits for particles of various size ranges and chemical compositions. These limits were incorporated into *Design Considerations for Datacom Equipment Centers*, Second Edition (ASHRAE 2009a). However, the NEBS limits do not indicate particle concentrations at which datacom equipment has been shown to fail. Rather, the requirements represent expected indoor concentrations when the datacom equipment center and equipment are located within densely populated urban environments that typically have high outdoor air pollution levels. The indoor concentrations were estimated using a material balance model, which assumed ventilation rates and filter efficiencies that may have been common within telephone switching centers. The contaminant levels have not been updated for at least 15 years although outdoor concentrations have been significantly reduced (Herrlin 2007).

Table 3.1 shows the average yearly levels of certain particulate contaminants based on predictive modeling and the 95th percentile values for these contaminants in the outdoor urban environment. The total suspended particulate (TSP) amount of airborne particulates is the sum of coarse particles (diameters $> 2.5\ \mu m$

Table 3.1 Average Annual Levels of Indoor Contaminants

Contaminants	Concentration
Airborne total suspended particulates (TSPs)	$20\ \mu g/m^3$ (7.22546×10^{-13} lb/in.^{3})
Coarse particles	$< 10\ \mu g/m^3$ (3.61273×10^{-13} lb/in.^{3})
Fine particles	$15\ \mu g/m^3$ (5.41909×10^{-13} lb/in.^{3})
Water-soluble salts	$10\ \mu g/m^3$ max–total (3.61273×10^{-13} lb/in.^{3})
Sulfate	$10\ \mu g/m^3$ (3.61273×10^{-13} lb/in.^{3})
Nitrites	$5\ \mu g/m^3$ (1.80636×10^{-13} lb/in.^{3})

[9.84252 × 10^{-5} in.]) and fine particles (diameters ≤ 2.5 μm [9.84252 × 10^{-5} in.]). Fine particles are composed of water-soluble salts, sulfates, and nitrates. The predictive modeling considers a building with a 10% ASHRAE dust spot rating, continuous operation of air-conditioning unit fans, and 10% outdoor air (90% recirculated air) (Bellcore 1995).

3.2.2   IEC 60721-3-3

IEC 60721-3-3 (IEC 2002) consists of multiple sections that define typical environments that equipment may be exposed to in various conditions. The standard also classifies groups of environmental parameters and the severities to which products are subjected when mounted for stationary use (permanently or temporarily) at weather-protected locations under specific use conditions, including periods of erection work, down time, maintenance, and repair. The standard is based on information acquired from actual field measurements worldwide. Section 3.3 of IEC 60721-3-3 covers the likely range of contaminants and other environmental parameters in a number of typical weather-protected environments ranging from controlled datacom equipment center environments to crude structures with little more than a roof. Environmental conditions directly related to explosion hazards, fire extinguishing, and ionizing radiation are excluded. Only those environmental conditions listed in Table 3.2 are considered. No special explanations of the effects on the products for these conditions are discussed in the standard.

Table 3.2   Environmental Parameters of IEC 60721-3-3 (IEC 2002)

Code	Parameter	Covers
K	Climatic conditions	Temperature, humidity, temperature change rate, air pressure, solar radiation, heat radiation, air movement, condensation, wind-driven precipitation, water from sources other than rain, and formation of ice
Z	Special climatic conditions	Modifies climatic conditions of a particular class combination
B	Biologically active conditions	Flora (e.g., mold and fungus) Fauna (e.g., rodents and termites)
C	Chemically active conditions	Sea salts and pollutants (sulfur dioxide, hydrogen sulfide, chlorine, hydrogen chloride, hydrogen fluoride, ammonia, ozone, and nitrogen oxides)
S	Mechanically active conditions	Sand and dust
M	Mechanical conditions	Stationary vibration and nonstationary vibration

The standard defines the usage environment by breaking it down into environmental parameters. Table 3.2 shows a list of the specific parameters covered in the standard.

The list in Table 3.2 is very comprehensive. Each of these conditions has a range of severities with corresponding measurable levels. IEC 60721-3-3 (IEC 2002) contains a set of tables specifying the severity levels typically associated with various real-world situations. The severities specified are those that will have a low probability of being exceeded.

Specifying a set of environmental class conditions brings about levels for each of the classes in Table 3.2. This includes specifying not only contaminants but also ambient environmental specifications.

The process used by manufacturers to specify a complete environment is fairly straightforward:

- Determine the type of environment in which the product will operate, following the suggestions in IEC 60721-3-3 (IEC 2002).
- Find the corresponding set of environmental conditions.
- Make exceptions as needed to suit the product requirements.

3.2.3   Federal Standard 209E-100,000 (M6.5) and ISO 14644-1 Class 8

Recent datacom workers entering the profession may not appreciate that there is a standard for datacom equipment center cleanliness. In years past, it had been an industry practice to classify cleanrooms according to *U.S. Federal Standard 290E-100,000, Airborne Particulate Cleanliness Classes in Cleanrooms and Clean Zones* (Institute of Environmental Sciences 1992). However, FED-STD-209E was officially withdrawn in November 2001 and was superseded by ISO 14644-1 (ISO 1999) and *ISO 14644-2, Cleanrooms and Associated Controlled Environments—Part 2: Specifications for Testing and Monitoring to Prove Continued Compliance with ISO 14644-1* (ISO 2000). The ISO 14644 series of standards (14644-1 to 14644-8) was established with various classifications for air cleanliness as well as methods for testing compliance. The ISO 14644 series of standards also includes an introduction to cleanroom design/construction/start-up considerations and operations. The ISO 14644 replacement standard classification differs from FED-STD-209E classifications as shown in Table 3.3. The two standards are approximately equal, except ISO 14644-1 uses new class designations, uses a metric measure of air volume, and adds three additional classes. The other new ISO 146442-2 standard provides requirements for monitoring a cleanroom or clean zone and requirements that act as verification of its continued compliance with ISO 14644-1.

ISO 14644-1 (ISO 1999) has become the dominant, worldwide standard for classifying the cleanliness of air. This standard gives the airborne particle limits to

Table 3.3   Airborne Particulate Cleanliness Class Comparison

ISO 14644-1 (ISO 1999)	FED-STD-290E (Institute of Environmental Sciences 1992)	
ISO class	U.S. English Class	U.S. Metric Class
1		
2		
3	1	M 1.5
4	10	M 2.5
5	100	M 3.5
6	1000	M 4.5
7	10,000	M 5.5
8	100,000	M 6.5
9	—	—

classify cleanrooms and associated controlled environments, such as datacom equipment centers, exclusively in terms of concentration of airborne particles. Examples of controlled environments include not only cleanrooms but also datacom equipment centers, telecom/datacom closets or /s, telecom/datacom switching centers, cable head ends, and cellular communication shacks and hubs. Table 3.4 provides maximum concentration levels for each ISO class. For a given particulate size, each successively higher classification allows approximately ten times as many particles as the previous class. Table 3.4 also shows that the ratio of particles of a given size remains generally constant for all classes. Work has now started on a revision to this standard, which is expected to be completed in 2009.

The ISO classes in Table 3.4 are based on the following formula:

$$C_n = 10^N (0.1/D)^{1.08} \qquad (1)$$

where

C_n = maximum permitted number of particles per m^3 ($in.^3$) equal to or greater than the specified particle size, rounded to a whole number

N = ISO class number, which must be a multiple of 0.1 and be 9 or less

D = particle size in μm (in.)

Most datacom equipment centers need to be kept clean so they are able to meet the cleanliness requirements of ISO Class 8 shown in Table 3.4 (Ortiz 2006). Class 8 allows 3.52 million 0.5 μm particles/m^3 (1.9685×10^{-6} in. particles/$in.^3$). Class limits must be met with the strictness of the 95% upper confidence limit. Chapter 5 of this publication discusses contamination testing and analysis.

Table 3.4   ISO Air Cleanliness Classifications
vs. Maximum Particle Concentrations Allowed (particles/m^3 [in.^{3}])

ISO Class	Maximum Number of Particles in Air (Particles in Each Cubic Meter [Cubic Inch] Equal to or Greater than the Specified Size) Particle Size					
	> 0.1 μm (3.93701 × 10^{-6} in.)	> 0.2 μm (7.87402 × 10^{-6} in.)	> 0.3 μm (1.1811 × 10^{-5} in.)	> 0.5 μm (1.9685 × 10^{-5} in.)	> 1 μm (3.93701 × 10^{-5} in.)	> 5 μm (0.00019685 in.)
Class 1	10	2				
Class 2	100	24	10	4		
Class 3	1000	237	102	35	8	
Class 4	10,000	2370	1020	352	83	
Class 5	100,000	23,700	10,200	3520	832	29
Class 6	1,000,000	237,000	102,000	35,200	8320	293
Class 7				352,000	83,200	2930
Class 8				3,520,000	832,000	29,300
Class 9					8,320,000	293,000

Note: Uncertainties related to the measurement process require that data with no more than three significant figures be used in determining the classification level.

3.3   PUBLISHED GUIDELINES AND LIMITS FOR GASEOUS CONTAMINANTS

Established gaseous composition environmental limits, listed in Table 3.6, have been published in standards such as IEC 60721-3-3 (IEC 2002) and GR-63-CORE (Telecordia 2006), which were developed for telephone switching centers and the equipment producer's own internal standards. The standard that most applies to airborne contamination in datacom environments is the International Society of Automation's *ANSI/ISA S71.04-1985, Environmental Conditions for Process Measurements and Control Systems: Airborne Contaminants* (ISA 1985). ISA is an international, nonprofit, technical organization that fosters advancements in technology for measurement and control in a wide variety of applications. Although ANSI/ISA S71.04-1985 classifies the level of airborne contaminants that are safe for electronics equipment, it also includes airborne gaseous contaminants. The standard establishes environmental corrosion levels by measuring the rate of corrosion buildup (i.e., corrosion thickness over time measured in Angstroms) using copper coupon strips and then quantifying the severity into one of four classes: G1, G2, G3, or GX (as shown in Table 3.5).

Table 3.5   ISA Corrosion Class Levels (ISA 1985)

ISA Class	Level	Description
G1	Mild	Corrosion is not a factor in determining equipment reliability
G2	Moderate	Corrosion is measurable and may be an issue in five years
G3	Harsh	It is probable that corrosion will occur within five years
GX	Severe	Only specifically designed and packaged equipment will survive

Datacom equipment centers located in urban areas have reported electronic equipment failures of lead-free restriction of hazardous substances (RoHS) components in the presence of small amounts of atmospheric sulfur and chlorine. These occurrences have prompted ISA to undertake a revision to ISA S71.04-1985 (ISA 1985) to include silver corrosion in its severity levels (*Data Center Journal* 2008).

The limits in Table 3.6 serve as a guide for specifying telecommunications equipment center environmental cleanliness, but they are not useful for surveying the corrosives or predicting the failure rates of hardware in the telecom environment for two reasons. First, gaseous composition determination is not an easy task. Second, predicting the rate of corrosion from gaseous contamination composition is usually not a straightforward exercise.

An additional complication to determining corrosivity is the synergy between gases. For example, it has been demonstrated that hydrogen sulfide (H_2S) alone is relatively noncorrosive to silver when compared to the combination of hydrogen sulfide (H_2S) and nitrous oxide (N_2O), which is very corrosive to silver (Volpe and Peterson 1989). Correspondingly, neither sulfur dioxide (SO_2) nor nitrous oxide (N_2O) alone is corrosive to copper, but together they attack copper at a very rapid rate (Johansson 1985).

Although Table 3.6 can be used to provide some indication of the possible harmful effects of a number of common contaminants, the datacom environment needs a single set of limits, which will require considerable study and research. As the industry works toward a single set of limits, caveats or exceptions to generally accepted limits will exist. These exceptions will improve as the interactions of concentration, composition, and the thermal environment combine and become better understood along with their effects on the datacom equipment.

Table 3.6 Published Gaseous Contaminants for IT Equipment

Gas	IEC 60721-3-3 (IEC 2002)	GR-63-CORE (Telecordia 2006)	ISA S71.04-1985 (ISA 1985)	One Manufacturer's Internal Standard
Hydrogen sulfide (H_2S)	$10\ \mu g/m^3$ (3.61273×10^{-13} lb/in.^{3}) 7 ppb	$55\ \mu g/m^3$ (1.987×10^{-12} lb/in.^{3}) 40 ppb	$4\ \mu g/m^3$ (1.44509×10^{-13} lb/in.^{3}) 3 ppb	$3.2\ \mu g/m^3$ (1.15607×10^{-13} lb/in.^{3}) 2.3 ppb
Sulfur dioxide (SO_2)	$100\ \mu g/m^3$ (3.61273×10^{-12} lb/in.^{3}) 38 ppb	$131\ \mu g/m^3$ (4.73268×10^{-12} lb/in.^{3}) 50 ppb	$26\ \mu g/m^3$ (9.3931×10^{-13} lb/in.^{3}) 10 ppb	$100\ \mu g/m^3$ (3.61273×10^{-12} lb/in.^{3}) 38 ppb
Hydrogen chloride (HCl)	$100\ \mu g/m^3$ (3.61273×10^{-12} lb/in.^{3}) 67 ppb	$7\ \mu g/m^3$ (2.52891×10^{-13} lb/in.^{3}) 5 ppb*	—	$1.5\ \mu g/m^3$ (5.41909×10^{-14} lb/in.^{3}) 1 ppb
Chlorine (Cl_2)	$100\ \mu g/m^3$ (3.61273×10^{-12} lb/in.^{3}) 34 ppb	$14\ \mu g/m^3$ (5.05782×10^{-13} lb/in.^{3}) 5 ppb*	$3\ \mu g/m^3$ (1.08382×10^{-13} lb/in.^{3}) 1 ppb	—
Nitrogen oxides (NO_X)	—	700 ppb	50 ppb	$140\ \mu g/m^3$ (5.05782×10^{-12} lb/in.^{3})
Ozone (O_3)	$10\ \mu g/m^3$ (3.61273×10^{-13} lb/in.^{3}) 5 ppb	$245\ \mu g/m^3$ (8.85119×10^{-12} lb/in.^{3}) 125 ppb	$4\ \mu g/m^3$ (1.44509×10^{-13} lb/in.^{3}) 2 ppb	$98\ \mu g/m^3$ (3.54047×10^{-12} lb/in.^{3}) 50 ppb
Ammonia (NH_3)	$300\ \mu g/m^3$ (1.08382×10^{-11} lb/in.^{3}) 430 ppb	$348\ \mu g/m^3$ (1.25723×10^{-11} lb/in.^{3}) 500 ppb	$348\ \mu g/m^3$ (1.25723×10^{-11} lb/in.^{3}) 500 ppb	$115\ \mu g/m^3$ (4.15464×10^{-12} lb/in.^{3}) 165 ppb
Volatile organics (C_XH_X)	—	$5000\ \mu g/m^3$ (1.80636×10^{-10} lb/in.^{3}) 1200 ppb	—	—

* Total HCl and Cl_2.

4

Prevention and Control

4.1   INTRODUCTION

This chapter suggests ways to limit the impact of particulate and gaseous contamination through prevention and control practices. Particulate and gaseous contamination has the ability to degrade every level of datacom equipment from the most basic electronic components to the most sophisticated information technology, infrastructure, and peripheral equipment. The datacom equipment contamination concerns and control methodologies discussed in this chapter are fitting for every size and level of datacom equipment and facilities. The ideas discussed in this section may not be applicable or appropriate for consideration within every datacom equipment environment. The reason is, no two facilities are exactly alike. Even identical physical facilities will experience different levels of contamination effects because of their different geographic locations.

Moreover, not all facilities require the same level of contamination control. The economic effects of equipment downtime in an enterprise-level facility are presumably much more significant than the downtime economic effects of a point-of-sale server located within a retail store. The economic risks of equipment downtime attributed to particulate and gaseous contamination will drive which level of contamination control is required within any datacom equipment installation.

Appendix A provides a basic framework for readers and for datacom equipment center facility managers responsible for prevention and control programs to separate their datacom equipment environments based on potential exposure to particulate and gaseous contamination.

Potential contamination sources exist everywhere and are unavoidable in any building, structure, or datacom environment. Contaminants in the form of gases, solids, and liquids may be hazardous to IT equipment operation if not adequately considered and managed. In short, prevention and minimization of equipment-threatening contaminants is fundamental to the long-term reliability and availability of datacom equipment. While there are many items to consider, strategies can be put in place to efficiently and effectively minimize the threat of most contaminants. Prior to design and construction, site selection is critical for long-term survival of IT equipment in datacom environments. Some sites may require extra equipment and systems to manage environmental hazards. For example, locations close to the ocean can have high airborne salt levels due to sea spray. Sea spray, which contains a high concentration of chloride ions, is largely responsible for corrosion of metal near the coastline. During the design and construction phase of a datacom environment, materials selection of everything from the raised-floor systems to the ceiling should be chosen to minimize contamination. Decisions for environmental pressurization, introduction of outside air, and finishes must also be made carefully to achieve the goal of minimizing threatening contaminants in the datacom environment. Infrastructure equipment selection, installation, maintenance, and physical placement of IT equipment all have an influence on the long-term cleanliness of the datacom environment once it is in daily operation. Operational strategies and policies must be clear, understood, and practiced by all datacom personnel and service providers to limit contaminants in the datacom environment. Even the best designed and constructed facility will become degraded if these policies are not included in the facility's standard operating procedure.

4.2 PREVENTION

The best way to control environmental and equipment contamination is by prevention. Preventing contamination from entering the datacom environment minimizes the need for contamination management. When contamination is introduced into the datacom environment, the difficulty of eliminating it increases dramatically. Contamination in the environment can quickly penetrate datacom equipment and, once infiltrated, the expenditure to remove it can be significant in terms of service and labor costs or even larger when hardware replacement is necessary. However, the biggest contamination consequences are equipment downtime and lost revenue.

4.2.1 Risk Assessment

There are a multitude of potential contaminants and countless ways in which datacom equipment can be impacted. It is important to note that not all datacom equipment center environments are equally susceptible to all contaminants. Datacom operators need to realistically acknowledge when a datacom equipment center environment may be susceptible. For instance, two facilities of identical design and construction may be impacted differently simply because of different geographic locations. Conversely, two facilities in the same geographic area may be impacted differently because of facility design differences. A risk assessment should be conducted to characterize the contaminant potential. The results of the assessment should feed into other planning activities so that action plans can be developed to mitigate the impact of contamination-related risks.

4.2.2 Facility Location

Datacom environment site selection, including identification of surrounding internal and external hazards, is an important consideration. When considering the location of a new or relocated datacom equipment center, the selection process should take into account contamination exposure risks from neighboring properties. These risks may include agricultural, chemical, biological, nuclear, and manufacturing processes; storage; and waste treatment operations. Facility location selectors also need to consider geographical locations that are prone to floods, tornadoes, volcanoes, or other acts of nature. Datacom equipment center contamination can also result from commercial transportation in proximity to the facility, such as heavy truck or train traffic and possibly airport operations with aircraft flight paths overhead. All of these sources of potential contaminants can put the datacom environment at risk. In reality, a completely risk-free datacom equipment center location is very difficult to find. Site selection usually involves compromise.

Protection from significant weather events, while costly, is an important consideration that connects both location and design. In flood-prone areas, it may be necessary to design all datacom equipment center areas with support equipment to be installed on upper floors in case the first level floods. In areas prone to tornadoes and hurricanes, it may be important to properly orient the building and to adjust the facility design to survive high winds.

Even in nonstorm situations, the physical location of a building can significantly impact the level of contamination within the datacom equipment center. The level of particulate matter (PM) (airborne or otherwise) outside a facility can and will find its way inside via cracks in the building, makeup air, and other exchanges with materials and people entering the datacom equipment center.

Urban areas contain man-made pollutants such as petrochemical materials from car tires, soot from combustion equipment, dust and debris from construction activity, etc. Rural datacom equipment center locations may experience more dust from loose soils and vegetation. Vacant lots and open spaces can suddenly become construction sites, exacerbating contamination problems. In both urban and rural settings, changes in wind direction and pressure will have an impact on the level of airborne PM and the level of infiltration inside the datacom equipment center.

Typically, urban locations are more at risk for vibratory events, but both urban and rural locations are susceptible. Vibration can be caused by roadway traffic, trains, airplanes, and nearby construction. Vibration can cause various building components to shift and either dislodge PM or create new PM.

4.2.3 Computer Room Design

The datacom environment should be located within a building away from potential local hazards and physically separate from space continuously occupied by humans. For example, operations such as cafeterias or toilets located above or boiler rooms, steam pipes, or parking garages located adjacent to or below datacom environments can negatively affect the datacom equipment center environment contamination level, especially in the event of an accident, act of vandalism, or act of terrorism.

The most effective method of dealing with contaminants is to keep them out of the datacom equipment center. Therefore, strict datacom equipment center practices should be established to minimize contaminants carried in by everyday occurrences. Absolutely no food or drink should be allowed in the datacom equipment center environment at any time. Crumbs from food or spilled liquids put datacom equipment at risk. Simply allowing food in the data center creates a careless attitude toward other contaminants. Cardboard boxes and IT equipment manuals should remain outside of the datacom equipment center in a designated location. Paper is another particulate source and is also a fuel source in the event of a fire. It is essential for facility owners and operators to designate an equipment unpacking and staging area to support moving in/out IT equipment. Tacky mats or contamination control mats should be installed just beyond the entrance(s) to the datacom equipment center to remove debris from shoes. There should be a set procedure for changing the mats so these remain effective.

Once a datacom equipment center facility location is determined, the physical design of the datacom environment can impact the amount of PM. Any opening can allow PM into the facility; thus, the location of openings is important. Exterior windows should be avoided in the datacom equipment center not only to minimize the infiltration of contaminants but also to minimize security threats and solar gain. If the no- exterior-windows guideline proves impossible, windows should be located on the leeward side of the building, particularly in high-wind areas. Doorways to the datacom environment should be isolated from exterior doorways in an attempt to form an airlock. At no time should a building exterior door be open while the datacom equipment center entrance door is open. Positive pressurization by the use of conditioned makeup air, air showers, or vestibules are possible options for reducing the influx of contaminants.

4.2.3.1   Attached/Adjacent Staging Areas

Areas outside the datacom environment should be provided to allow equipment to be delivered, unpacked, and staged prior to being installed in the datacom environment. Such an area allows equipment to be cleaned and prepared. It also enables crating and packing materials to be removed without contaminating the datacom environment. This area should be physically separate from the datacom environment, and air exchange should be prevented.

4.2.3.2   Attached/Adjacent Storage Areas

Areas outside the datacom environment should be provided for the storage of spare parts, equipment, and supplies. Keep materials loaded with PM such as old computers, cable reels, cardboard boxes, paper, or other materials capable of generating PM outside the datacom environment to control PM. It is also important to move parts and equipment through an area where they can be cleaned prior to being introduced into the datacom equipment center from storage.

4.2.3.3   Traffic Flow

The datacom environment should be designed with efficient foot and pushcart travel in mind. Long delivery paths increase dirt pickup before equipment enters the datacom equipment center room. If equipment is covered with plastic during transport, the plastic should be removed before the equipment is taken into the datacom equipment center room. Plastic accumulates PM through static adhesion; the PM can then be easily released into the environment of the datacom equipment center room. PM can also be released into the room when plastic covering is agitated during the removal process. Wheels of transport devices should be rolled over tacky mats prior to entering the datacom equipment center room. The datacom environment should be designed as a destination location within the facility structure, not a path or short-cut for people and materials to reach other locations.

4.2.3.4   Office and Operations Areas

A large quantity of PM can be generated and transported by people. Limiting the number of people working in the datacom environment can effectively reduce contamination. All datacom operators should have a desk or work area outside the datacom equipment center environment. The datacom equipment or network operations areas should be physically separate from the datacom environment and should be negatively pressurized relative to the datacom equipment center room if there is a direct doorway path between them. Only those people who need to physically interact with the datacom equipment should be in the datacom equipment center.

4.2.4   Computer Room Construction

Like most construction projects, a variety of materials are available to design and construct a datacom environment. It is very easy to overlook the PM characteristics of these materials among the countless other project details. Not all materials selected for use within datacom equipment center construction will be ideal from a PM viewpoint. Generally, there are trade-offs between choosing good PM materials and other considerations such as safety and cost. When considering datacom equipment center room construction, think of the value of the computer hardware itself and the information stored inside. Subsequently, think of the datacom equipment center having similar long-term value as that of an art museum. Contamination control for the datacom equipment center is just as important a design consideration as power, cooling, security, etc.

4.2.4.1 Wall, Ceiling, Underfloor Materials, and Surfaces

Construction materials and surface finishes that are good from a PM viewpoint are those that will produce no or comparatively little PM such as metal, glass, and plastic materials. While these materials are used for various components, they are typically not used for wall, ceiling, or floor coverings because of cost and other limitations. Most datacom equipment center facilities are constructed utilizing concrete floors, drywall walls, and cellulose suspended ceiling tiles. Unfortunately, each one of these materials generates significant PM, particularly during datacom equipment center construction. PM will occur after material installation as the result of either vibratory or abrasion events.

Underfloor Seal or Coating. Concrete materials (i.e., blended cements) are the workhorse of modern construction and are the base for almost every datacom environment. Ironically, concrete materials are also a potential source of dangerous contaminants. Exposed concrete materials continually oxidize and the surface breaks down. Concrete surface breakdown creates loose contamination consisting of sand and lime PM. Lime dust is particularly corrosive when mixed with water or humidified. Concrete surfaces must be protected against oxidation breakdown by sealing. Ideally, the seal is applied before the raised-access floor is installed, but a sealant can also be applied in an existing datacom facility. Even in an existing facility, some protection is better than no protection. Select a water-based, volatile organic compound (VOC)-compliant sealer that is intended for datacom equipment center applications. Most new concrete materials are treated with a curing agent to help harden the concrete and produce a better surface. These curing agents are often called *surface sealers* but are predisposed to sink into the concrete rather than remain on the surface as a surface protector.

A simple evaluation can be performed to determine the existence or condition of a concrete surface sealant. This is particularly useful for acceptance testing of a new datacom facility. To carry out this examination procedure, the examiner should put on laboratory glasses and hand protection at a minimum. Below is the process:

- Using plain water, clean a 152.4 mm (6 in.) diameter circle on the concrete surfaces to be evaluated to remove any surface contamination.
- Apply two to three drops of muriatic acid on the concrete surface in the center of the circle.
- If the muriatic acid remains on the concrete surface and looks like water, the concrete is adequately sealed for dust encapsulation purposes.
- If the muriatic acid reacts with the concrete by producing clear or yellowish foam, the concrete is NOT adequately sealed. Do not breathe the vapor produced by the reaction.
- Clean up the muriatic acid from the concrete surface test area, and properly dispose of the hand protection gloves as well as the clean-up wipes.

NOTE: Muriatic acid is dangerous if not properly used. Do not pour more than two to three drops of acid on the concrete floor surface. Unsealed concrete and large quantities of acid will produce a large reaction and dangerous quantities of fumes.

Wall Coverings. Most datacom equipment center environments are constructed using painted drywall (also called *gypsum board* or *wall board*) partition walls. This is a good material choice, provided the drywall surface is properly prepared and high-quality paint is applied. This will ensure that the surface will not chalk or rub off. Care must also be taken to ensure all cut drywall edges are properly covered or sealed (e.g., around outlets and other surface penetrations). Unused or abandoned holes and penetrations should be patched and painted during the same period of time.

Other types of wall construction and coverings have been used over the years. Porous surfaces such as fabric are not recommended, as they may capture and yield contaminants from the basic material. Such surfaces are also difficult to clean. All wall coverings and surface finishes must comply with applicable local building and fire codes.

Ceiling Tiles and Space. Ceiling tiles can be another significant contributor to datacom environment contamination. Most commercial lay-in ceiling panels are not suitable for use in datacom equipment center environments, since they are made from compressed cellulose and are highly friable, meaning that cellulose is easily broken into small fragments. Simple movements of the panels, either intentional or because of building vibration, can cause the edges of the panels to chip and break.

Acceptable panels are those with smooth surfaces that have wrapped or encapsulated edges. These types of panels are commonly used in food service kitchen and preparation areas. Just as the concrete in the underfloor space must be sealed to prevent surface oxidation and liberation of concrete dust into the airstream, any exposed concrete in the ceiling space must be sealed as well. Additionally, sprayed-on fire insulation, commonly used to protect structural steel, is a source of PM and should be sealed if present in the ceiling space.

Zinc Whiskers. Besides metal shavings and rust particles, the most common electrically conductive contaminants found in a raised-access floor datacom equipment center environment are zinc whiskers. Zinc whisker growth has been documented on a broad range of zinc-coated datacom equipment and raised-access floor support products. These include the underside of zinc-plated raised-access floor panels, support structures (pedestals and stringers), and associated mechanical assembly hardware. Wood core and concrete panels with flat steel bottoms are most susceptible. The steel surfaces may be finished with zinc by using either a hot-dip galvanization process or by using the electropassivation electroplating process. Hot-dip galvanized steel typically has a spangled (glittery) or mottled (spotted) appearance similar to the finished surface of a tin bucket. This method of plating is generally considered to be immune to whisker growth. However, electroplated zinc typically has a uniformly dull gray appearance and will develop crystalline growths (whiskers) of pure zinc perpendicular to the panel's plated surface. Whisker impact on computer equipment comes when these small metallic particles, typically 0.5–1.0 mm (0.02–0.04 in.) in length, are dislodged from their source and introduced into the datacom equipment through the equipment's air intake paths. This condition may result in a short circuit or unwanted bridging between conductors. While all exposed electronic circuitry is vulnerable, power supplies tend to fail in a dramatic fashion. Zinc whisker contamination may produce audible popping sounds when these whiskers arc across high-voltage or high-current conductors.

It is important to recognize that almost all electroplated zinc surfaces are potential whisker sources. Zinc whiskers have been found on a wide range of products inside operating datacom environments, including:

- raised-access floor panels, pedestals, stringers, and pedestal heads
- steel building studs
- suspended ceiling T-grid components and hanger wires
- concealed spleen ceiling grid components
- thin wall electrical conduit
- datacom equipment racks and cabinets
- datacom equipment cases and enclosures

Tin Whiskers. Similar to zinc whiskers, tin whiskers are tiny, electrically conductive, pure-metal, hair-like crystalline structures that grow from components and products that have electroplated tin as a final surface finish. Tin whiskers can grow in abundance and cause bridging and shorting between electrical conductors as well as component terminations. Exposed leads on electronic components are commonly plated with tin or a mixture of tin-lead to prevent corrosion and enhance solderability. As more manufacturers move toward restriction of hazardous substances (RoHS) compliance, manufactures are required to remove lead elements from the plating formula. Today, most lead-free electronic component terminations are plated using a pure tin plating process. Pure tin plating is a common source of whisker growth. There have been many documented findings of tin whisker growth on electronic components that has caused failures to occur inside the datacom equipment. A good source of whisker information can be found at the U.S. National Aeronautics and Space Administration (NASA) Web site, http://nepp.nasa.gov/whisker/.

The cause of both tin and zinc whisker growth mechanism(s) is not known or well understood. Many component manufacturers and plating industry professionals have developed effective ways to work around them by modifying the plating solution with other materials. Nonpure metals tend to be less susceptible to whisker growth.

4.2.4.2 Fit and Finish

Improperly fitted and finished interior systems can contribute to excessive contamination in the datacom equipment center. The amount of air moving through building cracks can be significant. With the unwanted airflow comes contamination either from outside the building or from contaminated areas within the building. This is one reason the datacom equipment center should be positively pressurized relative to the surrounding areas. Air-conditioning systems typically are not designed to maintain positive pressurization in the face of numerous or large air leaks. Although not the focus of this publication, such cracks and associated air leaks reduce operational efficiency by bleeding conditioned air from the datacom equipment center and therefore are covered generally here.

Areas of significant concern include:

- Ceiling—should be sized to fit snugly at the sides and ends within the grid. Broken or chipped panels will allow exterior building air to infiltrate the datacom center.

- Drywall—should be adequately caulked to the slab (at the base) and to the roof or slab of the adjacent floor. Consideration should be given to building movement and expansion as well as to fire rating when selecting material and caulk.

- Columns—can generate significant amounts of airflow because of the chimney effect. This airflow can carry contaminants into the datacom equipment center. Figure 4.1 shows a common occurrence where a large opening is created to facilitate the utility installation and is never repaired.

Figure 1.1 Column with a large hole.

4.2.5   HVAC System

The datacom equipment center HVAC system can be considered part of the datacom equipment center room construction. The HVAC system can be a significant source of both PM and gaseous contamination.

4.2.5.1   Makeup Air

Fresh makeup air is a building code requirement for installations with human occupants. The typical commercial HVAC installation has minimal filtration of the outside air before it is mixed into the return airstream, conditioned, and supplied back into the environment. It is not unusual for the same equipment design to be employed for the datacom equipment center environment makeup air system. However, if datacom environmental HVAC systems do not properly filter the outside air, the datacom equipment could become contaminated with all of the outside air pollutants. Lack of adequate filtration can be disastrous to datacom and infrastructure equipment if the outside air is heavily polluted. Outside air in all geographic locations has its own peculiarities relative to pollution. Countries without pollution standards are more at risk when using makeup air. Local or regional authoritative sources should be sought for up-to-date information about outside ambient conditions. For example, the U.S. Environmental Protection Agency (EPA) has a Web site that produces reports and maps of air pollution for any locale with the United States (www.epa.gov/air/data/). Also, the Federal Emergency Management Agency (FEMA) has a Mapping Information Platform that can provide hazard information with respect to volcanoes, forest fires, and wind storms (https://hazards.fema.gov/femaportal/wps/portal).

4.2.5.2   Positive Pressurization

Positive pressurization helps prevent contaminated air from entering the datacom equipment center environment. The use of positive pressurization with outside air not only keeps particulate contaminants out of the datacom equipment center, but it is also used to control corrosive gases and VOCs (Krzyzanowski and Reagor1991). Even though most installations in typical commercial business and clean industrial locations have an adequate quality of surrounding air, any air entering the datacom equipment center should be conditioned and filtered to ensure that datacom equipment temperature, humidity, and cleanliness stay within the datacom equipment specifications.

4.2.5.3   Humidification Systems

Thermal Guidelines for Data Processing Environments, Second Edition (ASHRAE 2009b) has updated the recommended environmental envelope for datacome equipment to recommend that the relative humidity in a Class 1 datacom equipment center be controlled with a lower-bound dew point of 5.5°C (42°F) and an upper-bound dew point of 15°C (59°F), or 60% relative humidity. ASHRAE Environmental Class 1 is defined as a datacom equipment center with tightly controlled environmental parameters (e.g., dew point, temperature, and relative humidity) for mission-critical operations (ASHRAE 2009). Most facilities achieve the necessary humidification through specialty humidifiers. (See Chapter 21 of the *2008 ASHRAE Handbook—HVAC Systems and Equipment* [ASHRAE 2008] to see the range of equipment types.) While several technologies exist, all of them have trades-offs between water quality, maintenance, and contamination. Because standard tap water is rarely suitable for use in humidification systems, care is required in matching the water quality and associated treatment equipment to the selected humidifier. Water contains particles, bacteria, dissolved solids, and a wide range of chemicals. These impurities can affect the humidification system and its performance. Many of the impurities in the water remain when the water is converted to vapor. The issue becomes one of whether the impurities are removed in the purification steps or deposited in the humidification equipment itself, requiring a higher frequency of maintenance.

There are five categories of water that could be used to supply water for the humidifier:

1. Potable water essentially has no treatment other than filtration. Unless the potable water source is one of very low dissolved solids, this source will significantly increase maintenance and repair costs.

2. Softened water is a viable source for humidification. One of the main challenges is the removal of hardness (e.g., calcium carbonate, etc.) from the water that would otherwise produce mineral deposits or scaling. In a water softener, an ion exchange resin typically uses salt to regenerate the resin, which allows the exchange of sodium for calcium. The higher levels of sodium in the water will not cause scaling in the system and are generally not problematic in the airstream. Care must be taken that the operation of the softener includes appropriate rinsing, as the regenerant chemical is salt, and the chloride used can be harmful if left in the water and carried over in a mist. In addition, the softener does not appreciably reduce the total dissolved solids in the water, which results in residue in the humidifier. Water softeners are reasonably easy to run, and costs are low. Water softeners may be the ideal choice for smaller data centers.

3. Reverse osmosis removes a majority of water-borne contaminants by applying a pressurized water stream into a membrane. The cleaner product water is provided on the outlet of the system while the contaminants are carried away, concentrated in a reject stream of water. Reverse osmosis is capable of rejecting the majority of contaminants and will provide suitable feed water to the humidifier. Reverse osmosis results in a minimum of humidifier maintenance or cleaning. The downside to reverse osmosis is that there is higher water usage because of the reject stream. The higher water use needs to be balanced against the chemical use of softeners and deionization systems. From an environmental perspective, the end user needs to decide which they prefer to minimize: water use or chemical use.

4. Deionization uses ion exchange resin in the same way as a softener but removes both anions and cations—as well as provides a high-purity water stream (a softener simply exchanges sodium for hardness cations, such as calcium and magnesium, without markedly improving the water purity). Deionization systems are typically the most complex but will provide the highest water quality. Very pure deionized water is an aggressive fluid. When exposed to air through the humidifier and delivered to the room, the water rapidly loses its aggressive nature. While in the piping system, there is potential risk to some materials. The humidification supplier should be consulted about material compatibility if deionized water is used. An additional source of information is *Liquid Cooling Guidelines for Datacom Equipment Centers* (ASHRAE 2006).

5. Boiler feed water systems exist in many larger facilities. This water could be used for humidification systems as it generally has a reduced level of contaminants; however, the specific water chemistry needs to be fully understood and compared to the requirements of the humidification system.

Any water-treatment or humidifier system does better when it is operated regularly. When the water is stagnant, biological contamination and settling of suspended solids can be a problem. It is best to rinse the system prior to use and to implement the humidifier manufacturer's recommended cleaning, operation, and maintenance practices.

Particles can be a variety of sizes and need to be filtered out of the water. The humidification system manufacturer's guidance should provide the level of filtration. As each type of humidifier has different sensitivities to particulate contamination, no general guideline can be provided. However, consideration of desired particle sizes for airborne contaminants may be a guide to selecting the water filtration level.

The extent of water conversion to steam or water vapor needs to be looked at carefully. If the system generates a water mist, the size of the droplets and the ability of the HVAC system to fully evaporate them before entering the IT equipment are critical. The subject of hygroscopic dust is discussed at length in Chapter 2, Section 2.4.1.5. Consult this section for more information. A humidifier that generates a fine spray (instead of a fog) may create undesired airborne contaminants, especially if the water has a significant quantity of dissolved solids or PM.

The sizing of the water softener, reverse osmosis, or deionization system is dependent on both the water flow rate and the amount of total dissolved solids in the feed water. The specific manufacturer's guidance should be followed.

This book specifically excludes the issue of personal health and comfort associated with biological contamination in the humidifier. See the ASHRAE Handbook chapter on humidifiers (ASHRAE 2008) for further reference and guidance. Unfortunately, the purification processes may actually remove the bacterial control chemistry, and as such, the system design may need the review of appropriate environmental hygiene experts.

In all cases, the best solution is to look at the total cost of ownership of the entire system, humidifier, and water treatment equipment. Additional first costs in the water treatment system typically allow for longer periods of time between maintenance activities and longer life of the humidification system. Also, there must be a comparison between the required water quality costs and the energy operating expense. Different types of humidification systems require different levels of water quality and conversion energy for the liquid-to-gas phase.

Humidity can impact the rate of corrosion in the datacom environment as the corrosion process produces PM (e.g., rust and other particulates). In general, conductive anodic filament growth (Montoya 2002), deliquesce of hydroscopic salts, or condensation can occur on computer components if the humidity level is too high.

Humidity control can be challenging in free air-cooling applications such as facilities with significant air-side economizer installations. External ambient humidity considerations should be included when designing humidification control for these facilities.

4.2.5.4 Air Filtration

Filtration is an integral part of any air movement device that may be part of the equipment, added to an existing system, or a stand-alone unit (e.g., electrostatic). Filtration and air cleaning removes unwanted PM and gaseous materials from the airflow paths. Some level of air filtering is necessary in datacom equipment center environments. Filtering may take place in recirculated air using the computer room air-conditioning (CRAC) units, in makeup air supplied from the outside environment, and in some datacom equipment. In HVAC applications, this involves air filtration and in some cases, air cleaning (for gas and vapor removal).

The ASHRAE filter standard *ANSI/ASHRAE Standard 52.2-2007, Method of Testing General Ventilation Air-Cleaning Devices for Removal Efficiency by Particle Size* (ASHRAE 2007a) is currently used to rate filters based on their collection efficiency, pressure drop, and particulate-holding capacity. This standard measures arrestance, dust spot efficiency, and dust holding capacity. Arrestance is a measure of a filter's ability to capture a mass fraction of coarse dust, and dust spot efficiency is the ability to capture particles within a given size range. ASHRAE Standard 52.2 also measures particle size efficiency expressed as a minimum efficiency reporting value (MERV) between 1 and 20. Table 4.1 shows the values from ASHRAE Standard 52.2. Particulate filters with a MERV rating of 8 are recommended as a minimum in *ANSI/ ASHRAE Standard 127-2007, Method of Testing for Rating Computer and Data Processing Room Unitary Air Conditioners* (ASHRAE 2007b) and are commonly available for CRAC units. A MERV 11 or 13 filter should be used with air-side economizers or makeup air units. It is worth noting that *ANSI/ ASHRAE Standard 62.1-2007, Ventilation for Acceptable Indoor Air Quality* (ASHRAE 2007c) recommends at least a MERV 6 filter upstream of all cooling coils and, in outside air units located in areas where PM10 is exceeded.

Higher-efficiency filtration may be necessary for some datacom equipment center environment installations. Filtration always impedes the airflow through the air-conditioning system as well as through datacom and infrastructure equipment; therefore, it is important for facility and equipment designs to take the filter impedance into account. Filters should be inspected, replaced, or cleaned at regularly defined intervals to minimize airflow impedance. Some pieces of IT equipment use active alarming based on differential pressure drop to indicate when filters need to be serviced. Manufacturer-defined time intervals may not

Table 4.1   Values from ASHRAE Standard 52.2

	ASHRAE 52.2 (ASHRAE 2007a)					
MERV	3–10 μm	1–3 μm	0.3–1 μm	Arrestance	Dust Spot	Dust Spot
1	< 20%	—	—	< 65%	< 20%	
2	< 20%	—	—	65%–70%	< 20%	> 10 μm
3	< 20%	—	—	70%–75%	< 20%	
4	< 20%	—	—	> 75%	< 20%	
5	20%–35%	—	—	80%–85%	< 20%	
6	35%–50%	—	—	> 90%	< 20%	3–10 μm
7	50%–70%	—	—	> 90%	20%–25%	
8	< 70%	—	—	> 95%	25%–30%	
9	< 85%	< 50%	—	> 95%	40%–45%	
10	< 85%	50%–65%	—	> 95%	50%–55%	1–3 μm
11	< 85%	65%–80%	—	> 98%	60%–65%	
12	> 90%	> 80%	—	> 98%	70%–75%	
13	> 90%	> 90%	< 75%	> 98%	80%–90%	
14	> 90%	> 90%	75%–85%	> 98%	90%–95%	0.3–1 μm
15	> 90%	> 90%	85%–95%	> 98%	~95%	
16	> 95%	> 95%	> 95%	> 98%	> 95%	
17*	> 99%	> 99%	> 99%	—	> 99%	
18*	> 99%	> 99%	> 99%	—	> 99%	0.3–1 μm
19*	> 99%	> 99%	> 99%	—	> 99%	
20*	> 99%	> 99%	> 99%	—	> 99%	

* Filters viruses and carbon dust.
I-P Conversions:
$3–10 \ \mu m = 1.18 \times 10^{-4}$–$3.94 \times 10^{-4}$ in., $1–3 \ \mu m = 3.94 \times 10^{-5}$–$1.18 \times 10^{-4}$ in., $0.3–1 \ \mu m = 1.18 \times 10^{-5}$–$3.94 \times 10^{-5}$ in.

be accurate because of wide variations in the severity of datacom equipment center environmental conditions. The existence of significant concentrations of VOCs may require activated charcoal or permanganate-based filters to reduce VOCs to an acceptable environmental level.

4.2.6 Fire Suppression System

The most commonly used fire suppression systems for datacom equipment center environments are:

- water based—sprinklers (NFPA 2007) and mist suppression (NFPA 2006)
- gaseous based—clean-agent gas suppression (NFPA 2008) and halon gas suppression (NFPA 2009)

Foams, dry and wet chemical agents, and inert gases are less commonly found in datacom equipment centers; therefore, their effect on computer equipment is not considered. The NFPA sources listed in this book's References and Bibliography section (2006, 2007, 2008, 2009) suggest that clean-agent and halon-type suppression systems do not have a negative impact on computer equipment. However, fire suppression systems should be evaluated to determine if the agents themselves, in the event of an accidental discharge, or the by-products that form when exposed to excessive heat or fire can harm the computer equipment.

4.2.7 Mechanical Malfunction

CRAC and computer room air handler (CRAH) equipment can also produce contamination internally. Fan units, bearings, and pulleys can become misaligned and rapidly deteriorate during normal operation. Fan belts frequently degrade or disintegrate first when moving parts become misaligned.

There are three primary causes of excessive drive belt wear: the geometry of the actual belt design itself and materials used to fabricate it, drive belt alignment within the equipment's application, and the drive belt tension of its application. In a study that will be presented during the 2010 ASHRAE Winter Conference (Stack 2010), three original equipment manufacturer (OEM) drive belt geometries (raw edge, seamless, and wrap-molded designs) and materials of high-end specification grade designs are reviewed. It is shown that raw-edge belt design performed better than the seamless design. The wrap-molded belt also performed better than the seamless, but not as well as the raw-edge belt. The study indicates that when non-OEM drive belts are used as OEM replacements, these belts have significantly shorter lives due to excessive wear and stretching. It is recommended that OEM drive belt replacements be used because they are usually designed and machined to match as a set of two or three for uniform belt tension. This matching was reported to ensure exact common dimensions and characteristics to ensure balanced loading.

Proper drive belt alignment is also critical. Proper alignment means the belt is positioned exactly 90 degrees to the shafts being driven. Drive belt misalignment will cause side loading on the belt or imbalance between paired belts if more than one belt is used, both of which result in excessive belt heating. The alignment of the belt is impacted not only by the original factory design but also by adjustments made in the field to vary the blower speed. This is especially true if variable-pitch pulleys are used. As a variable-pitch pulley is adjusted, the center distance between the pulleys is changed as well as the center line of the pulley. That is because one side of the pulley is fixed in place while the other side is moved in and out to adjust how far the belt drops into the pulley groove. If possible, variable-pitch pulleys should be changed to fixed dimension units once the desired dimension is known in the field. In addition, as the system runs, the drive belt heats up. The heat can cause the drive belt to lose tension, stretch, and slip if the center distance of the pulleys is not properly adjusted. Once a belt begins to slip, it heats up faster, causing more slip and wear.

The heat also causes the drive belt to harden. This will result in belt cracking and loose particles. Reduced belt wear greatly minimizes the concern of belt dust that can negatively impact IT equipment hardware. The airflow through the equipment will push belt particulate contamination into the datacom environment. Most CRAC and CRAH units are filtered on the air inlet or return side, and PM produced inside the unit is downstream of the filter media. CRACs and CRAHs should be periodically inspected and repaired, as needed, to prevent equipment failure and PM contamination. Drive belts must be maintained and precisely aligned at all times. The use of variable-frequency drives or electronically commutated motors may eliminate the use of drive belts and associated potential PM contamination.

4.2.7.1 Ceiling Returns

Historically, datacom environments were designed as a typical office environment. Suspended ceilings were installed to enclose unsightly mechanical systems and to provide a pleasing aesthetic appearance. To respond to cooling challenges, many new datacom environments are constructed without suspended ceilings. This design approach allows hot air to rise farther away from the datacom equipment on its return back to the cooling equipment. A number of datacom equipment center owners are simulating this airflow pattern by taking advantage of the ceiling space above the suspended tiles. This approach is done by installing open diffusers over hot equipment aisles and then ducting the hot air from the ceiling space back to the cooling unit returns.

This airflow implementation presents new challenges for controlling contaminants. The ceiling space is now an active airflow return area and is subjected to contamination accumulation. All active airflow returns collect PM contamination and require periodic cleaning to limit their impact within the datacom equipment center environment. Cleaning the ceiling space can be a difficult procedure due to the nature of the area, difficulty of access, and density of components and surfaces on which contamination can accumulate. Finally, extreme caution must be exercised when cleaning this space, since it is directly over active datacom equipment.

4.2.8 Operational Procedures

After a datacom equipment center facility is designed and constructed, there are a variety of operational procedures that should be adopted to assist datacom operators in preventing the entry and impact of contamination within the datacom equipment center environment.

4.2.8.1 Record Keeping

Record keeping may seem like a simple task, but keeping track of various datacom equipment center activities in and around the datacom environment will provide the foundation for many decisions. Most management decisions are based on data. Without data, decisions are random and unreliable. The ability to identify the impact of various activities is important. Frequently, there is a direct correlation among a number of activities in or around a datacom equipment center and contamination-related problems.

It is crucial that equipment anomalies and failures are recorded. Subtleties and trends can only be detected if there is recorded data to analyze. Too often, datacom operators disregard failures as random events and neglect to record the event because of the failure's limited effect. In reality, the failure may be part of a larger failure trend that could have been detected if the records had been updated and maintained.

4.2.8.2 Control Access

Datacom workers are another source and contributor of PM within the datacom equipment center. Limiting the number of people entering the datacom equipment center is an excellent way to limit the PM contributed by them. The following steps should be considered to reduce the number of people in the datacom equipment center:

- Install a datacom equipment center access control system for physical admittance into the space. Numerous people access the datacom equipment center simply because they are unrestricted.
- Access should be restricted to those individuals, facility maintenance personnel, and service providers necessary to support the datacom equipment and environmental infrastructure.
- Periodically review access security control records to ensure that only those people who are required for datacom equipment center operations are granted access. It is not uncommon for employees and service providers to remain on approved access lists long after their assignment has changed or access is no longer required.
- Review service and work processes and change procedures to allow only those tasks absolutely necessary to be provided inside the datacom equipment

center. Many tasks that historically required physical interaction with the equipment can now be accomplished remotely. Encourage staff to interact with and manage systems remotely.

- Consider strategically locating datacom equipment and people within the datacom equipment center to a concentrated area. By grouping such activities and equipment, the need for people to move throughout the datacom equipment center is greatly reduced. The concentrated area should be located as far as possible from the most sensitive datacom equipment.
- Establish an access control audit team and ensure that their requirements are addressed in the design process.

4.2.8.3   Track-Off Matting and Contamination Control Mats

People and materials represent a noteworthy contributor of PM in the datacom environment. People track PM on clothing, footwear, and portable equipment. The majority of materials are wheeled into the datacom equipment center on carts, dollies, or hand trucks. A combination of track-off matting and contamination control mats (tacky mats) can significantly reduce PM from people and materials.

Track-off mats are typically constructed using man-made fibers affixed to a rubber backing. These mats are commonly placed inside doorways to remove soil and to absorb moisture from footwear. Track-off mats with different fiber densities and naps are available for a large assortment of contaminant conditions. Track-off mats require maintenance and can be serviced and reused. When servicing mats, the product manufacturer's recommended procedures should be followed. Professional service companies offer contamination mat rentals that include periodic exchange and other contamination control cleaning services.

Contamination control mats, also known as *tacky mats* or *sticky mats*, are available as both renewable and disposable products. Both tacky and sticky mats feature a mildly sticky surface that captures contaminants that are lightly bound to the host's contact surface (e.g., dust on footwear and moving cart wheels). Typically, renewable matting is permanently glued down. Vacuuming and damp-mopping the surface on a daily or more frequent basis renews the surface. Disposable mats are constructed using multiple layers of thin adhesive plastic sheets. When the exposed layer becomes soiled, it is peeled away and discarded to expose a fresh layer. Effective decontamination needs at least six footfalls, three for each foot or three full wheel rotations. Using an average person's walking stride, the matting length required is approximately 4.6 m (15 linear ft). Ideally, this distance would be applied to all contamination control mats. In practice, this is not always possible due to space constraints.

4.2.8.4   Datacom Equipment Center Change Control

Many datacom equipment center disasters have been caused by poorly planned or poorly timed datacom activities. Adopting and enforcing change control processes and logically questioning all aspects of the internal procedures can help ensure that maintenance activities are reviewed and evaluated for potential datacom equipment center impacts. For instance, what are the PM risks connected to using a ladder to open a ceiling tile above a datacom equipment rack? What additional contamination risks are associated with the type of work that will be done above the ceiling? Who in the datacom equipment center organization will assess the risk of the contamination that may be produced around the rack? Who will be notified if a problem occurs? These are the kind of questions that should be answered and resolved before any datacom equipment center change activity is allowed to proceed.

4.3 CONTROL

Despite efforts to prevent the ingress of PM and gaseous contamination into the datacom equipment center, some contamination is inevitable. The following areas speak to the importance of contamination controls and offer guidance.

4.3.1 Monitoring

Monitoring is the first step to controlling contamination. Using some type of monitoring system to evaluate and chronicle the condition of the datacom environment is helpful for determining whether existing prevention or control activities are adequate and effective.

Visual observations and physical inspections are best and least expensive monitoring systems available. First, examine surfaces in the datacom equipment center room where PM may accumulate, just as you would look for dust and dirt in your home—especially in hard-to-reach places. Look closely at built-in filters and air intake paths within the datacom equipment center. If the datacom equipment center has a suspended ceiling, carefully remove selected ceiling tiles and examine the topsides for PM. After any datacom or facility work is completed, inspect the work area for dirt, dust, wire clippings, metal shavings, and other types of PM. If PM rapidly appears or is observed to accumulate faster than expected, there is reason to investigate. Unexpected accumulation or detection of PM is frequently the result of equipment failure or the breakdown of a control or separation barrier. Both occurrences are serious and should be corrected. For instance, consider the possibility of CRAC drive belt deterioration, which has been previously discussed in this chapter. Appropriate response and correction is necessary to limit PM impact for continual operation of the datacom equipment.

4.3.2 Equipment Failures—Severe

Commercial datacom and infrastructure equipment are some of the most robustly designed products. Datacom equipment failures are infrequent, but when they do fail, they generally occur because of a lack of maintenance or environmental operating conditions. Monitoring symptoms and causes of equipment failures is an excellent method of identifying underlying problems. Datacom equipment center operators should practice environmental monitoring, analysis, and risk management within the standard operating procedure. Most contamination service providers can also provide monitoring and failure analysis services.

4.3.3   Equipment Failures—Nonsevere

Many contamination-related failures first appear as intermittent failures. Tracking intermittent abnormalities, which requires datacom equipment shutdowns and/or reboots, is an excellent method of identifying problems. Collecting and aggregating incidence reports from multiple failures can identify contaminant-connected trends.

Once PM is determined to be the reason of the failure, it is valuable to understand the type and source of the PM.

4.3.4   Periodic Maintenance Plan

Similar to other locations within a facility, the datacom equipment center must be maintained and kept clean. Even with the restrictive procedures identified previously, PM will enter the datacom equipment center. The scope and frequency of the maintenance activities will depend on the amount of activity within the datacom equipment center. Datacom equipment centers with limited or no physical barriers to contamination will require more frequent cleaning than those that have controls. Datacom equipment centers with frequent equipment upgrades and rearrangements and the personnel required to complete the work tend to generate more PM and will necessitate more frequent cleaning than static, lights-out datacom equipment centers.

Due to the sensitive nature of the environment, great care is required for all cleaning activities. Contractor experience is an important consideration, as improper equipment selection or improper usage can result in serious damage to the flooring system and/or datacom equipment center. Most cleaning-related contamination problems can be attributed to poorly trained workers. Be sure to hire a professional and reliable cleaning company that specializes in datacom equipment center cleaning services. Always ask for references.

Datacom equipment center cleanliness can be maintained by establishing a consistent cleaning schedule. Cleaning frequency should be increased during construction or other contamination-producing activities. The essential areas to clean and time intervals are described below.

Underfloor Volume

Minimum of once per year:

*   Remove large PM that cannot be vacuumed by hand.

- Vacuum all accessible surfaces with a high efficiency particulate air (HEPA) filter vacuum.
- Use mechanical actions (i.e., wiping) to remove PM that cannot be removed by vacuuming alone.

Floor Surfaces

Minimum of once per week:

- HEPA vacuum floor surface.
- Do not dry mop or sweep the floor surface. This does not remove contaminants and can resuspend settled material.

Minimum of once per quarter:

- Damp-clean (i.e., mop) entire floor surface.
- Scrub floor surface as needed.
- Remove PM from beneath equipment racks and cabinets by brushing and HEPA vacuuming.

NOTE: Most raised-floor panels are covered with high-pressure laminate (HPL). This surface material is very durable but should not be abused. High-speed buffers should not be used. Low-speed floor scrubbers can be used. Water and non-corrosive cleaning solutions can be used sparingly and should be removed promptly from the surface. HPL flooring should never be waxed. Some, but not all, HPL materials are conductive for electrostatic discharge (ESD) control. Coating this surface with floor finish (e.g., wax) blocks the ESD path through the panel. Vinyl composition tile (VCT) floors should be periodically stripped and refinished with antistatic, conductive floor finish.

Equipment and Cabinet Exteriors

Minimum of once per quarter:

- HEPA vacuum and/or damp-wipe surfaces.

Environment—Walls, Sills, Ledges, etc.

Minimum of once per quarter:

- HEPA vacuum and/or damp-wipe surfaces.

NOTE: Cleaning should never be done with compressed air inside the datacom equipment center. This spreads and scatters PM into the datacom equipment center environment where most resettles on or in other equipment. If the use of compressed air is considered a practical and correct way to clean a piece of hardware, the hardware should be moved outside the datacom center environment beforehand. Before using compressed air, it should first be determined that this cleaning method will not force PM further into the equipment, perhaps into areas that are even more difficult to see or clean. Compressed air can also pose a static electricity danger due to the velocity of the air passing through hosing and nozzles. Ionizing air guns and nozzles are commercially available to eliminate static from the airstream when required. Be sure compressed air lines have proper dryers installed. Do not use compressed air to clean equipment while air-conditioning power is connected.

4.3.5   Nonroutine Events

4.3.5.1   Clean Up Your Mess!

Any controlled environmental condition can quickly deteriorate. A relatively contaminant-free area can unexpectedly become contaminated as a result of equipment moves and rearrangements, equipment installation or repair, facility maintenance, etc. The contamination control strategy and procedure should include requirements and instructions to clean up after any of these procedures. This may mean using in-house staff or an outside datacom cleaning service to do the work. Someone has to clean up the mess. Moreover, PM cleanup can be dangerous if left unattended. The longer contamination is left unrestricted, the greater the potential for the PM to become airborne and get inside the datacom and infrastructure equipment.

4.3.5.2   Contractor Cleanup

The best in-house contamination process and controls can quickly unravel at the hands of untrained or unaware contractors. Service providers and contractors should be held responsible for limiting the amount of contamination they generate and for cleaning up the contamination created. The cleaning method and procedures should be consistent with the established cleaning protocol.

4.3.5.3   Contamination Control During Construction and Other Major Events

Some of the most common construction activities inside a working data center are drilling holes and cutting wire. Drilling operations should be done outside the data center whenever possible. When drilling inside the datacom equipment center, a powerful vacuum with a HEPA filter should be used next to the drill bit to capture and remove as much chip or dust PM as possible before it scatters into the environment. The area around the drill site should be masked and covered to the maximum extent possible to contain debris that the vacuum misses. If it becomes necessary to drill into the concrete floor under a raised-access floor used for air delivery, the drill site should first be surrounded with air dams to prevent the airstream from blowing and scattering the concrete dust. Drilling operations should not start until an above-floor hand vacuum appliance has been activated for use around the drill site. For some drilling work, such as core drilling, the work can be done wet. Wet drilling processes significantly minimize dust, but the amount of water applied must be controlled. The resulting slurry must be carefully and completely removed when the procedure is completed, or else the debris dries and becomes dust.

When major construction work occurs outside the datacom equipment center, particularly in a nearby or adjacent space, it becomes necessary to re-examine all of the datacom equipment center environmental protection systems to ensure that these can control the greater amount and types of PM that will be created. Improved filtering may be needed, provided that the filters don't overly restrict airflow. Filters may need to be changed more frequently than normal. Walls should be examined for damaged sealing, separation, penetrations, and cracking. If any of these conditions exist, they should be repaired before construction work begins. Tacky mats should be installed at additional locations outside the data center and changed more regularly. Thought should also be given to require contractors, at their expense, to clean up areas surrounding the datacom equipment center on a regular basis to remove construction dust before it can enter the datacom equipment center environment. For major and prolonged construction work, professional cleaning of the datacom equipment center may even be needed several times during construction. The extra cleaning and PM mitigation costs should be expected and made a part of the construction bid specifications. If construction is taking place above the datacom equipment center, thought should be given to the possibility that PM may be released through vibration of the building's ceiling components. Extra safety measures should be taken by the contractor to avoid and prevent any possibility of water leakage or smoke contamination from welding operations into the datacom equipment center.

4.3.5.4 Disaster Response and Contamination Control

Datacom equipment centers are the heart and nerve center of most companies today. Unfortunately, even the best-protected and best-managed facilities can have their datacom equipment centers compromised by environmental contamination or other unexpected disasters. The best way to protect a datacom equipment center is to create a strategic disaster recovery plan and to test it. If an unexpected event occurs, the plan can be immediately placed into action to allow the datacom equipment center to return to operation as quickly as possible with high reliability.

Datacom owners, facility operators, and plan designers should partner with qualified external resources that can quickly help restore datacom equipment center operations. Specialized datacom equipment center disaster recovery providers offer a range of services that may include recovery strategies, damage assessment, cleaning, equipment restoration, and media data recovery. The majority of disaster recovery and support providers supply specially trained on-site response teams within hours of a disaster.

There is no industry standard or agency certification to ensure that datacom equipment center environmental service providers are trained and competent. Datacom planners should survey and clearly understand what services are being offered and how they compare with known industry best practices. When selecting a data-com equipment center service provider during the disaster recovery planning stage, choose only those companies that have fully trained technicians who understand and follow ISO 14644-1 (ISO 1999) for critical environments, as discussed in Chapter 3. Once disaster plans are developed, established, and tested, it is critical that all datacom equipment center employees and support personal understand, follow, and practice contamination control procedures. Developing a strategic datacom equip-ment center disaster plan is outside the scope of this document.

4.4 SPECIAL CONSIDERATIONS BASED ON DATACOM EQUIPMENT CENTER LEVELS

In general, a wide variety of datacom environments exist and each environment will have different exposures to PM and gaseous contamination. Today, there are no adopted datacom industry standard datacom equipment center contamination level classifications. A proposed datacom equipment center classification level standard is outlined in Appendix A. The proposed standard for each level is based on observed conditions. The attributes assigned to each level are intended to serve as a baseline for that level. Datacom equipment center owners and operators should use the proposed classification levels to compare and evaluate their own facilities along with the other areas discussed in this chapter.

5

Contamination Testing and Analysis

5.1 INTRODUCTION

The objective of datacom equipment center contamination analysis is to determine the presence and sources of contamination that may have damaging effects on the reliability of computers.

A datacom equipment center contamination analysis should begin with a visual inspection of the relative amounts and types of contaminants as well as their potential sources. While not a quantitative measurement, visual inspection is important for getting an overall idea about the contamination in a datacom equipment center to help plan a course of action for contamination analysis. A datacom equipment center survey helps determine where to place the monitoring devices to measure the airborne particle count, total suspended particles (TSPs), and volatile organic compounds (VOCs), as well as where to collect the settled dust specimens. The main elements of a site visual inspection and general audit are:

- environmental history of the data center
- ventilation system and layout of the data center
- choice of sampling locations

The environmental history of a datacom equipment center includes the history of datacom equipment and infrastructure hardware failures, temperature and humidity records, and complaints about odors. Oftentimes, the root cause failure analysis of field-returned hardware may identify a chemical species, gas, or settled dust that interacted with the printed circuit boards and may be the cause of the failures. The datacom equipment center visual survey helps identify the potential sources of the gaseous or particulate contamination.

The ventilation system and layout of the datacom equipment center should be mapped to document the airflow pattern in the datacom facility. The locations of all computer hardware, tape drives, disk drives, printers, administrative areas, air diffusers, modular air-conditioning units (MACUs), and HVAC units should be included on the layout drawing. There are several notable areas to examine specific to the HVAC system:

- Air distribution scheme for the building—should be mapped out and should include the level of air filtration, the areas with shared air return, the degree of outdoor air used, and the locations of the outdoor-air intakes.
- Air filters—should be inspected to verify that they are being well maintained. Inspection can include direct visual observation of dust loading and/or measuring the pressure drop across the filters.
- Air humidifiers—can release salts dissolved in water used within the humidifier into the environmental air leaving the humidifier. These salts can cause problems such as electrical shorts on printed circuit boards.

Once the history and the layout of the datacom equipment center have been obtained and documented, it is necessary to choose the locations where contamination monitors should be placed and dust collected.

- Air monitoring devices should be placed:
  - immediately downstream of filters, MACUs, HVACs, and humidifiers
  - at the entry doors to the data center
  - in the spaces between the administrative areas and datacom equipment
  - in areas around the datacom equipment
  - in areas around the printers

- Dust samples should be collected from:
  - internal surfaces of the datacom equipment
  - internal surfaces of humidifiers
  - under raised-access floor surfaces and the floor support grids and stanchions
  - internal surfaces of ductwork near the datacom equipment

Sampling equipment and procedures for air monitoring and dust sampling measurements are addressed in the next sections.

5.2   AIRBORNE PARTICLE COUNTS

Airborne particle counters (APCs) count the number of particles in the air per unit volume. Some test equipment counters can also determine the particle size distribution by counting particles in various size ranges.

An APC is a relatively simple device. A pump brings air into a fixed volume at a constant rate for a certain amount of time. A laser beam illuminates the particles and light is redirected or absorbed. A detector measures the amount of light scattered to determine the particle quantity (Lighthouse 2006).

Particle counter pumps operate at a constant speed. The speed varies by manufacturer and model but is generally 0.1 cfm (0.00283 m^3/min) or 1 cfm (0.0283 m^3/min). Particle counters are able to differentiate particle size via their channels. Each channel can analyze a specific size range of particles such as 0.1, 0.3, 0.5, and 5 μm (3.93701 × 10^{-6}, 1.1811 × 10^{-5}, 1.9685 × 10^{-5}, and 0.00019685 in.). Counter models can have one to six channels or more. Some APCs can give data for each channel in real time so that all particle sizes can be monitored in real time. Other APCs give only an average of all particle sizes.

There are three main types of particle counters: handheld, portable, and remote. Handheld devices are small, portable devices that are convenient for taking a quick measurement at various areas of the datacom equipment center. While they do not always have the same functionality of a larger portable model, many handhelds can collect the same types of data. Handheld devices generally take readings at several-second intervals for a real-time analysis.

Portable devices are bigger than handheld devices and can be powered by a wall outlet or a rechargeable battery. Portable devices are very useful when an area needs to be monitored over a long period of time. Data can be collected and stored on the counter for analysis at a later date.

Remote particle counters allow for multipoint monitoring, with the collected data reported back to one central location. This is generally required for a cleanroom facility that needs constant monitoring on a real-time basis. Particle sensors are placed around the room and data is transmitted and collected centrally.

Particle count monitoring in a datacom equipment center is generally not something that is needed on a daily basis; it is usually done only when there is a notable problem that could be caused by contaminants. While a handheld device can be helpful for periodic site checkups, a portable device will work best for troubleshooting. If possible, choose a counter that can track several particle sizes so the data will include the average count for each size range available instead of one total average value.

Measurements should be taken one or two weeks at a time. If possible, take measurements throughout the year because weather and seasons can affect the particles in a datacom equipment center.

Take as many readings as possible throughout the data center. The following areas are recommended to characterize a datacenter (Krzyzanowski and Reagor 1991):

- Doorways—Take one to two sets of readings at each doorway.
- Human activity/high-traffic areas—Take one set of data in each area of major activity (e.g., desks, chairs, workstations, etc.).
- Under a diffuser—Take one set of readings for makeup air particle count.
- Air handling unit (AHU)—Take one set of readings at the inlet and one set at the outlet of a certain AHU. This will verify the differences between the upstream and downstream conditions and will detect filter problems.

- General data center characterization—Take one set of readings in at least three generic places in the data center. For large datacom equipment centers, divide the area into imaginary sections and take at least two measurements in two normal areas.
- Outside—If possible, take a measurement outdoors several times a day where the inlet makeup air is entering.

The outdoor count is important for normalizing data taken at different times of day or during different seasons (Krzyzanowski and Reagor 1991). The outside air can greatly affect the indoor air quality in certain datacom equipment center facilities. If there is a noticeable spike in indoor air particles, comparing the indoor data with outdoor data can help determine whether the particles are from an indoor or outdoor source. If the outdoor particle count has increased similarly to the indoor, it is likely that the indoor particles are the result of outdoor air entering the datacom equipment center. If the outdoor source has stayed the same while the indoor count has increased, it is more likely the problem is not caused by the outdoors.

After the data is collected, chart the particle size. Check for extremes in each size range. Compare data between each set of measurements taken, such as between different areas of the data center, different times of the year, etc. Are any extremes showing up between data sets? These extremes can provide clues as to where or when more testing needs to be done. Further steps can be taken to identify, or possibly prevent, a problem that appeared during the analysis.

5.3 TOTAL SUSPENDED PARTICULATES

Another way to examine airborne particles is by looking at the total amount of suspended particles in the air. To measure the amount of TSPs, use a photometer. Photometer units provide a concentration of particles in a given area, generally in mg/m^3 (lb/ft^3).

TSP photometers are similar to APCs in their operation. The light scattered off the particles is turned into a voltage and compared against a calibrated standard to give a value (Keady 2000). The photometer cannot measure individual particles but instead takes measurements from a cloud of particles. The photometer fails for dust particles that are greater than 10 microns, as those particles are too big for the typical device to read. Often, if the dust is not completely dispersed in the photometer and ends up in clumps, the amount of dust will appear to be less than the actual amount.

A test instrument also exists that can pump air onto a special adhesive disk. Once the particles are on the disk, a scanning electron microscope (SEM) analysis can be performed on the particles to determine their composition. This is most useful when there is a major contamination problem but the source is unknown. The major elements found from the analysis can be matched to other known sources such as outdoor dust, chemicals in humidifier water, concrete dust, or smoke.

5.4   MASS CONCENTRATION

Units are available to measure the mass concentration of particles. In the simplest form, a piece called an *impactor*—a cylinder tube with several removable plates spanning part of the diameter—sends air through a nozzle to increase its speed. The air is turned sharply just before it hits a perpendicularly placed plate. Large particles cannot follow the air line and are deposited on the plate. Smaller particles follow the air line and continue to another impactor plate stage. The stages are repeated for the particulate sizes required, such as PM10 and PM2.5. The dimensions of the impactor stage determine the size of particles collected. The deposited particles can be analyzed gravimetrically or chemically.

Also, a piezobalance can be used to directly measure and output a mass concentration of particles. At a high level, an airstream enters a single-stage impactor to trap particles of a specified size and larger. The smaller particles (i.e., those sizes that are desired for testing) move through a nozzle and become charged. A crystal at the bottom of the unit is made to oscillate. The charged particles collect on the crystal and increase its weight, changing the oscillating frequency. The change in frequency can be correlated to a mass concentration.

5.5   CORROSIVENESS OF PARTICULATE MATTER

Not all particulate matter (PM) is corrosive. Very few types of PM cause datacom hardware to corrode and fail. A method for determining the corrosiveness of PM is described step-by-step below.

- Wires are soldered to interdigitated cards using rosin mildly activated flux.
- Cards are inserted in a 40% relative humidity (RH) (50°C [122°F]) chamber and stabilized.
- The interdigitated comb areas on cards are sprinkled with PM/debris.
- 15 V is applied across comb coupons, and leakage current is plotted versus time.
- RH is raised every three days in 10% steps, and the RH at which the leakage current rises to reflect 100 MΩ surface insulation resistance is noted.
- The PM is deemed corrosive if the PM causes the comb area surface resistance to decrease below 100 MΩ in an environment with 90 RH compared to what is experienced in a datacom equipment center environment.

An example of the results from the step-by-step method is shown in Figure 5.1.

5.6   VOLATILE ORGANIC COMPOUNDS

Identifying the source of unknown odors in the datacom equipment center is one of the most difficult and expensive problems that environmental investigators encounter. VOCs are organic chemical compounds that have high enough vapor pressures under normal conditions to significantly vaporize and enter the

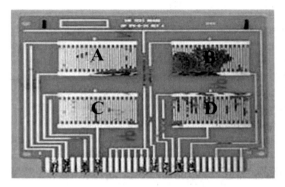

A = sheetrock dust
B = Poughkeepsie, NY, tap water residue
C = concrete dust
D = cement dust

Figure 5.1 Interdigitated card results.

atmosphere. VOCs originate from a broad range of sources in a datacom equipment center. VOCs are detected by first adsorbing them into charcoal filters and then analyzing the charcoal filters in a laboratory using gas chromatography/mass spectrometry (GC/MS). There are two common methods of collecting VOCs in charcoal filters. In one technique, air is pumped through the charcoal filters. In the other, a passive technique, no air is pumped through the charcoal filter. Instead, VOCs are adsorbed into the filter through diffusion over a longer period of time, typically a month. Passive monitors use no pumps. These techniques are only used in extreme situations where the source is unknown.

5.7   REAL-TIME GASEOUS MONITORING

Real-time monitoring consists of direct gas monitoring and real-time corrosion measurement. Direct gas monitoring can be done on a real-time basis with electronic devices that respond to changes in gas composition in a matter of minutes. Levels of many gases in the parts per billion (ppb) ranges can be detected. Today's available real-time monitors can detect ammonia, hydrogen chloride, hydrogen sulfide, nitrogen oxides, and ozone, sulfur dioxide, and total volatile organic compounds (TVOCs). Lower detection limits for the these gases are in the 0.1–1 ppb range, except for the TVOCs, which have a limit of 20 ppb.

A very convenient and quantitative way to determine the gaseous corrosivity of a datacom equipment center environment is the so-called "reactive monitoring" method described in *ISA S71.04-1985, Environmental Conditions for Process Measurements and Control Systems: Airborne Contaminants* (ISA 1985). Copper coupons are exposed to the environment for 30 days and quantitatively analyzed, using coulometric reduction, for corrosion film thickness and chemistry. Silver coupons can be included with copper coupons to gain a greater understanding into the chemistry of the sulfur-bearing gases in the environment. For example, sulfur dioxide alone will corrode silver to form silver sulfide (Ag_2S), whereas sulfur dioxide and hydrogen sulfide (when combined) will corrode both copper and silver, forming their respective sulfides.

There are four severity levels in the International Society of Automation (ISA) classification of reactive environments for corrosive gas contaminants (ISA 1985):

- Severity level G1, *Mild,* is defined when the corrosion film total thickness is less than 300 Angstroms in 30 days exposure. A G1 environment is sufficiently well controlled, such that the corrosion rate is too low to effect equipment reliability.
- Severity level G2, *Moderate,* is defined when the corrosion film total thickness is 300–1000 Angstroms in 30 days exposure. A G2 environment has a measurable corrosion rate that can degrade equipment reliability.
- Severity level G3, *Harsh,* is defined when the corrosion film total thickness is 1000–2000 Angstroms in 30 days exposure. A G3 environment has a high corrosion rate that may require environmental controls or specially designed and packaged equipment to avoid corrosion damage.
- Severity level GX, *Severe,* is defined when the corrosion film total thickness is greater than 2000 Angstroms in 30 days exposure. A GX environment has a high corrosion rate that may require environmental controls or specially designed and packaged equipment to avoid corrosion damage.

Although reactive monitoring gives the average corrosion rate over the 30-day period after the copper and silver coupons are exposed to the datacom equipment center environment, any short-term variations in the gaseous contamination levels are missed.

Real-time corrosion monitors use highly sensitive metal-plated quartz crystal microbalances (QCMs) to continuously measure the corrosivity of the environment. QCMs are microprocessor-controlled devices that can measure the total environmental corrosion attributable to airborne molecular contaminants. Commercially available monitors include copper and silver-plated QCMs that can measure real-time changes in corrosion rates as a result of the changes in gaseous pollutant levels. Real-time reactivity monitors are capable of providing information on corrosion levels as low as 1 ppb. Real-time monitoring allows preventive measures to be taken immediately, such as shutting off the outside air from entering the datacom equipment center should the outside air become too corrosive compared to the standard applied.

5.8   SETTLED DUST ANALYSIS

Identification of the chemistry of settled dust in a datacom equipment center environment serves two useful purposes. By examining the chemistry of the dust, one can identify its ionic characteristics and, as a result, its tendency to cause short circuits at different voltages on electronic circuit boards. Dust chemistry analysis also helps identify the dust's origin.

Settled dust analysis is accomplished by collecting the dust on a sticky tape stud, which is later examined using a SEM. The sticky tape is electrically conductive, which eliminates the need to carbon coat the dust particles that, being generally nonconductive, would otherwise electrically charge when exposed to the SEM electron beam. The dust on the sticky tape is examined in a SEM with energy-dispersive X-ray analysis (EDX). The EDX spectrum, which is displayed as counts versus X-ray energy, identifies the dust elements and provides a rough indication of their concentration. From the elements at hand and their approximate concentration, this assessment can provide an idea of the chemistry of the dust. For example, if the elements in the dust are mostly magnesium and chlorine, it can be concluded that the dust is most probably magnesium chloride ($MgCl_2$). Given that $MgCl_2$ adsorbs moisture from the air at a RH level as low as 35%, $MgCl_2$ will get wet and become ionic in a typical datacom equipment center environment with RH in the 45%–50% range. The ionic bridge between conductive features at different voltages will support metal migration and eventual metallic bridging of the features (i.e., electrical short-circuiting).

Settled dust analysis also helps to identify the source of the dust. For example, dust could originate from the concrete pavement outside the building and be carried into the datacom equipment center by personnel traffic, from human activity in or near the datacom equipment center, from soil near the air intake duct, or from water used to humidify the data center air. Analysis of the EDX elemental spectrum is compared to the elemental spectrum from the various suspected sources to identify the composition of the settled dust.

6

Air-Side Economizers

6.1 INTRODUCTION

Air-side economizers are used in cold and temperate climates to save energy in datacom equipment center facilities by means of using outside air as a technique for cooling the datacom equipment. Air-side equipment that typically includes a sensor and a filter allows outside air to enter the datacom equipment center when humidity, temperature, and other conditions reach a predetermined level. More standards information is sited in Section 6 of *ANSI/ASHRAE/IESNA Standard 90.1-2007, Energy Standard for Buildings Except Low-Rise Residential Buildings* (ASHRAE 2007d).

Integrating economizers for free cooling in datacom equipment centers is the subject of several other Datacom Series books, including *Best Practices for Datacom Facility Energy Efficiency*, Second Edition (ASHRAE 2009c), which outlines four concerns regarding using air-side economizers: increased particulate matter (PM), increased gaseous contamination, loss of humidity control, and temporary loss of temperature control.

6.2 IMPLEMENTING AIR-SIDE ECONOMIZERS

Datacom equipment centers in most climates can significantly benefit from an air-side economizer. Datacom equipment center cooling can often be served by using outdoor air during cooler weather, particularly at night when the outside air temperatures may be even cooler. Air-side economizer use can result in significant cost savings, depending on the size of the economizer load and the hours of operation. The main concern often expressed for not considering or using air-side economizers is that large amounts of outside PM and gaseous contaminants will enter and settle on datacom equipment and cause datacom equipment center equipment to degrade, wear out, or shut down. Second, datacom operators also fear loss of humidity control. There is some validity for these concerns, given that documented failures have occurred. Although this book does not cover issues related to the contamination and filtration of open water systems, it should be noted that a water-side economizer may also be used to provide free cooling. When the weather is cold enough, water is circulated through the cooling towers, which are, in effect, giant evaporative coolers. As the tower water is sprayed into the air, some of it evaporates, cooling the rest of the air down. A large heat exchanger transfers the cool water to the regular chilled-water loop. The heat exchanger prevents cross-contamination of the chilled-water loop with water that has been exposed to the air, airborne dust, etc., in the towers. In colder climates, closed-cell towers or dry coolers are utilized to eliminate the risk of contamination of the fluid. Since no contaminants are brought into the datacom equipment center, operators may choose to implement water-side economizers to eliminate the contamination concerns associated with air-side economizers.

Research completed by Lawrence Berkeley National Laboratory (LBNL) found that compounds called *hydroscopic salts*, when combined with relative humidity (RH) above 80%, are capable of impacting electronic equipment in ways that could lead to datacom equipment failures. However, properly designed, installed, and maintained filtration and humidity systems can minimize or keep out contaminants. Nonetheless, external contaminants entering into the datacom environment with the introduction of large amounts of outside air from air-side economizers may become a problem and must be mitigated as discussed in this section.

Even though the steady-state environmental conditions within the datacom environment are usually quite good, the introduction of contaminated outside air from an air-side economizer may result in intermittent malfunctions that can be catastrophic. Some sources of these malfunctions include the introduction of large quantities of PM from nearby agricultural activities and wind storms. These contaminants can block filters and restrict the required volume of outdoor input air, which can result in partial loss of cooling within the facility. Contaminants can also cause a buildup of PM on and in the datacom equipment, which can result in reliability and performance degradation attributable to higher internal temperatures or shorting of components from the deliquescence of the PM.

Soot and gaseous contamination from fire, volcanoes, natural vegetation, building materials, vehicles, hazardous material spills in transportation accidents, and building exhaust, as well as other unforeseen sources of contamination can cause unintentional signaling of the datacom equipment center fire detection and prevention system. Contaminant sources that could enter the datacom equipment center with an air-side economizer may confuse the fire detection system and result in the activation of cross-zone alarms. This type of activation may trigger the release of the datacom equipment center's fire suppression agent, shut down cooling fans, and result in an unscheduled system outage. The unscheduled outage can result from either the emergency power off (EPO) of datacom equipment center room power or from a loss of cooling that causes a rapid temperature increase that, when detected, automatically shuts down the datacom equipment to protect its internal components.

Large amounts of humidification added to offset dry air introduced into the datacom room by an air-side economizer or a large amount of latent cooling occurring on the cooling unit coils can release mineral salts into the air and then into the datacom equipment. These salts become conductive and corrosive at various RH levels. The introduction of the salts is exacerbated when water vapor is also introduced from large building humidification systems or local ultrasonic systems, unless de-ionized water is supplied (ASHRAE 2005). Both of these systems introduce water directly into the room where it is intended to evaporate rather than mixing water molecules into the air from an evaporative system like a steam generator or an infrared lamp over a pan of water. Regardless of which humidification system is used, water quality and purity are important in reducing scale buildup within the humidifier equipment and eliminating airstream mineral dusting.

Despite contamination risks, air-side economizers still represent a significant opportunity to save energy by eliminating the need for the mechanical refrigeration process when the external ambient conditions are suitable. Air-side economizers can certainly be utilized in a datacom environment and should be given full consideration during the design process, but care must be taken, with regard to contaminants, when implementing these systems. There are several things to consider in order to deploy air-side economizers successfully:

- **Location.** Understand the surrounding environment and identify potential threats. For example, is the datacom facility located near a hazardous industry plant (e.g., paper mill), a farm, or an airport?
- **Weather.** The wind direction is an important consideration in addition to having the proper temperature and humidity conditions for air-side economizer operation.
- **Nature.** Catastrophic events like forest fires or volcanic eruptions can send contaminants thousands of meters into the air, which can travel over long distances.
- **Filtration.** The selection of filters must include consideration of the amount of airflow and pressure drop necessary to sustain the needs of the datacom equipment. Filters can also be chosen that are treated to adsorb a specific impurity known to be present in the datacom equipment center ambient location. For example, activated charcoal is good for trapping carbon-based impurities as well as elements like chlorine.
- **Monitoring and Controls.** For datacom environments with air-side economizers, it is necessary to incorporate real-time monitoring and controls capable of reacting quickly to unexpected events that could release particulates or corrosive gases. Detection equipment in the marketplace today can provide real-time monitoring and reporting of corrosive, odorous, hazardous, and toxic gases and can gauge the real-time effects of corrosion on electronics. One such product uses patented copper- and silver-plated quartz crystal microbalance (QCM) sensors to measure the mass accumulation of corrosive film on sensitive metals. By applying the proper conversion factors, the product correlates the mass gain to

corrosion thickness, expressed in Angstroms. This measurement can then be related to the *ISA S71.04-1985, Environmental Conditions for Process Measurements and Control Systems: Airborne Contaminants* (ISA 1985), classes shown in Section 5.7 of Chapter 5. Something as simple as a transportation accident involving a vehicle carrying hazardous material can disburse unwanted contaminants and expose the datacom environment. Airflow or pressure drop monitoring must also be incorporated across the filters to sense the amount of blockage due to PM. Monitoring and control systems must be able to automatically detect a change in the environment before the datacom detection and protection system's alarm or the temperature in the computer room rises high enough to shut down equipment. The control system must automatically convert from an economizing cooling system to a supplemental (usually direct expansion [DX]) cooling system. On-site staff have to be notified and/or alarms must be sent to off-site monitoring support staff.

Statements to the effect of "the computer room has been running for years on outside air and nothing has happened" are equivalent to saying "I have been driving my car for 50,000 miles and have yet to change an air filter, yet it still runs." It is a question of when, not if, a disaster is going to happen. It is essential that the proper monitoring and controls are in place to react to a contamination situation that might shut down the datacom environment. Contamination in a datacom room can be devastating to the operation of the facility, but it does not have to be disastrous. With the proper filtering, monitoring, and controls systems in place, even intermittent events can be properly handled.

7

Summary

The tendency to focus on the physical structure of the datacom infrastructure (e.g., the power and cooling provisions) can distract datacom facility designers and operators from focusing on the significant threat to datacom equipment posed by particulate, gaseous, and chemical contamination. Natural and human-generated airborne contamination is ubiquitous in the exterior environment and can easily enter the datacom environment. Other contamination sources may exist within the datacom equipment center environment itself.

Datacom equipment can be affected by particulate and gaseous contamination in a number of ways (i.e., mechanically, chemically, or electrically). Equipment susceptibility to contamination is increasing as a result of the trend toward higher datacom equipment package density.

There are a number of existing standards that discuss particulate and gaseous contamination, but no one standard provides enough information for datacom operators to minimize operational risk. Each of the existing standards contains valuable information, but the scope and intent of each is different, so comparing them is difficult.

Certain active measures may reduce the effects of particulate and gaseous contamination. These start with the site selection for the datacom facility and include design of the facility and datacom room and airflow through the facility as well as other operational considerations. Filtration systems may provide a solution to contamination issues but only if filters are correctly designed and implemented. Operational policy considerations are also important in controlling contamination. For instance, housekeeping policies governing food and drink, cleaning, construction, printers, cardboard boxes, and so forth can reduce datacom contamination.

Periodic facility inspections can help identify risk to datacom equipment. Sample collection and analysis, along with particle counts, can further quantify potential threats. Environmental history and ventilation system layout are also important considerations. Sample coupons placed in the environment may be used to quantify corrosion threats.

The trend toward forced induction of exterior air for datacom cooling through the use of air-side economizers presents an increased risk of particulate and gaseous contamination in the datacom equipment center. Air-side economizers may reduce cooling costs but make filtration and humidity control more difficult.

It is important to consider particulate and gaseous contamination, in addition to the many other factors that must be balanced, to successfully operate a datacom environment.

As this book has discussed, particulate and gaseous contamination in datacom environments is multifaceted. To maintain a high level of IT equipment dependability and availability, it is critical to view contamination in a holistic way and to acknowledge that the datacom equipment center is a dynamic environment. There is no single remedy or panacea that will solve all particulate and gaseous contamination exposures or problems within the datacom equipment center environment. The most successful contamination control architect will be the one who can mobilize talent across many disciplines and integrate many of the ideas contained within this book (as well as new ideas that evolve over time) into his/her standard operating procedure.

Appendix A

Proposed Datacom Environment Contamination Levels

In general, a wide variety of datacom environments exist and each environment has a different exposure to particulate matter (PM) and gaseous contamination. A proposed datacom equipment center Level 0–4 contamination classification is outlined here. The proposed classification for each datacom equipment center level is based on observed conditions. The attributes assigned to each level are intended to serve as baselines for the levels.

The reader can use the proposed datacom environmental contamination levels to provide a common reference point and vocabulary for both equipment deployment and for the development of a contamination prevention and control program. For example, it would be reckless to deploy an enterprise-class IT datacom equipment device in a Level 0 environment and expect to achieve reliable operation. Different prevention and control protocols should be uniquely established for different level environments.

DATACOM ENVIRONMENT LEVEL 0

- Open area with no physical barrier to ambient conditions.
- Contains inactive datacom technology (e.g., terminations, punch down blocks, and patch panels).
- May contain some active datacom technology (e.g., hubs, switches, and routers).
- Contamination composition is the same as that of the surrounding area.

DATACOM ENVIRONMENT LEVEL 1

- Same as Level 0, but datacom technology is housed within a cabinet.
- Cabinets may have solid panel doors or mesh doors.
- Contamination may be moderately mitigated by filtering on the cabinet.

DATACOM ENVIRONMENT LEVEL 2—
SERVER CLOSET (BAILEY ET AL. 2007)

- Typically a riser or wiring closet.
- Enclosed environment with no special environmental considerations.
- Room or area is physically separated from adjacent areas with little security.
- Often contains inactive datacom technology (e.g., terminations, punch down blocks, and patch panels).
- May contain some active datacom technology (e.g., hubs, switches, and routers).
- May contain more mid-range or high-end computing and/or telecom technology.
- Contamination is dependent on the amount of airflow into/out of room.

DATACOM ENVIRONMENT LEVEL 3—
SERVER ROOM (BAILEY ET AL. 2007)

- Secondary computer location.
- Enclosed environment with upgraded security and possibly an uninterruptible power supply.
- Environment has air conditioning, but conditions aren't optimum.
- Rooms may or may not have a raised-access floor.
- Room contains all types of datacom technology.
- Contamination is dependent on room design and level of PM donor material transported into/out of the room.

DATACOM ENVIRONMENT LEVEL 4—
DATA CENTER (BAILEY ET AL. 2007)

- Localized, mid-tier, or enterprise class.
- Enclosed environment with expertly designed and operated environmental systems, as well as physical and/or digital security.
- Designed to function only as a data center.
- Maintained to manage all environmental conditions, including contamination.
- Contains all types of datacom technology.

Appendix B

Field Contamination Occurrences

The major repurcussions of contamination on datacom equipment center and computer equipment operation are often intermittent. Issues don't arise from minor irregularities that can influence the system's steady-state condition but occur when there is a severe change in one or more of the facility's systems.

For example, if floor panels are lifted in a contaminated datacom equipment center without the smoke detection system being deactivated first, single or multiple alarms may occur. Normally, raised-access floor datacom equipment center detection systems are cross-zoned using both photoelectric and ionization-type detectors. The cross-zoning configuration provides protection against false discharges of the suppression system by requiring one of each type of detector to be in the alarm state prior to the initiation of a suppressant discharge. However, severe contamination conditions and service provider activities may make these safeguards ineffective. Without immediate action by on-site facilities personnel, alarm activation could cause a release of the fire suppression agent, a shutdown of the air handlers, an activation of the computer hardware disconnecting means, or a combination of these events. Even if the datacom equipment is not shut down, some datacom hardware will not operate very long without proper cooling if the air handler systems are unavailable.

Following are some actual illustrations where contaminants have impacted the operation of the datacom equipment center.

- **Agricultural Field in Proximity to a Datacom Center.** The datacom equipment center was impacted by a nearby agriculture field that was plowed twice each year, in the spring and fall, during the time the datacom equipment center room was cooled using an air-side economizer. Numerous times, dust clouds generated from the field plowing went unnoticed and blocked the air intake filters of the economizer system. The economizer's compromised filter system caused an increase in the datacom equipment center facility and datacom equipment temperatures. In addition, the fine particles of the dust triggered the smoke detectors. If datacom facility personnel had not been on-site to react immediately, such as during nights and weekends, the datacom center may have experienced an unscheduled outage. Eventually, personnel realized that the economizer and facility had to be serviced and cleaned twice a year, resulting in an additional operating expense.

- **Datacom Center Affected by Low Relative Humidity because of Regional Wind Patterns (Santa Ana Winds).** This event occurred in a datacom equipment center that used an air-side economizer as much as possible and periodically experienced intermittent hardware failures. The root cause of these intermittent failures resulted from external wind pattern shifts. When strong, extremely dry, offshore winds occurred, the relative humidity (RH) in the datacom equipment center dropped as low as 10%. When this occurred at night, the air temperatures were low enough to maintain the use of air-side economizers, but the humidification system could not keep up with the introduction of the very dry air. With no one on-site to react to these dry conditions, the RH plummeted and the hardware failed.

- **A Pharmaceutical Lab with a Massive Engine-Generator Plant to Supply Emergency Power to the Facility.** At this facility, the generators were started and tested each week during normal business hours. During start-up and testing, a huge plume of smoke was released into the air, and the local winds carried the smoke to surrounding buildings. The fumes entered into the buildings and datacom center, activating multiple smoke alarms and forcing the evacuation of personnel. Fortunately, the maintenance staff was available during the day and aborted the dumping of the inert gas fire suppression system, thereby preventing a datacom outage. However, if this event had occurred at night or on a weekend, the datacom center may have experienced an unscheduled outage since no one would have been immediately available to react to the alarms.

- **Air Intake for the Datacom Center is Adjacent to a Parking Lot Serving a Neighboring Company's Loading Dock (Another Diesel Smoke Incident).** In this example, the datacom equipment center's air-side economizer air intake was located adjacent to a neighboring company's parking lot and loading dock. During normal operations, truck drivers would park in the lot and idle, waiting their turns at the loading dock. One night, multiple alarms were set off because of the amount of diesel exhaust fumes that entered the datacom center.

Appendix C

Future Work

Lawrence Berkeley National Laboratory, University of California Berkeley, and Pacific Gas and Electric Company are collaborating on a study of data center indoor air quality (IAQ) using an air-side economizer (Ganguly 2009). Although the results are not available for publication in this book, the following is information about the study as well as where to find the report.

C.1   OBJECTIVES

The objectives of this study follow.

1. Measure particulate concentration (size range: 0.01–2.5 μm [3.93701×10^{-7}– 9.8452×10^{-5} in.]) in a data center using partial and full air-side economizer modes with three different ratings—minumum efficiency reporting value (MERV) 7, 11, and 14 for HVAC filters.
2. Measure total mass concentration of sulfate, nitrate, total carbon, and elemental carbon using a filter setup. Compare the mass concentration values with changes in MERV filter ratings in full and partial economizer modes. Temperatures and relative humidity (RH) will also be measured.
3. Measure variation in fan energy and the overall energy savings from the economizer system in different MERV filter scenarios.

C.2   SAMPLING SITE

The sampling site is a data center in Sunnyvale, CA, located in an office building. The building is located within a mile of two high-traffic highways. It is a medium-sized facility (616 m^2 [6630.568817 ft^2] area, 2.7 m [8.858267717 ft] ceiling height). It has rows of tall (2 m [6.56167979 ft]) cabinets, with racks in which the server, data storage, and networking equipment are vertically stacked.

The cabinets are arranged in hot-aisle/cold-aisle configuration. Recirculation and mixing of air among aisles is prevented by using blanking panels in the racks and curtains at either end of the hot aisle. The data center utilizes air-side economizers, which are located along with the air-handling units (AHUs) in an adjacent room. Figure C.1 shows the layout of the data center and the AHU room. The data center does not have computer room air-conditioner (CRAC) units or a raised-access floor.

The cool outdoor air is mixed with return air from the data center and passes through a bank of MERV 7 filters. The energy management system (EMS) at the Sunnyvale site is set to provide 85% of outside air to the total flow when the outdoor temperature is below 21.1°C (70°F) and reduce it to 1% when it is above the setpoint. The data center goes through about 50 air changes per hour (ach).

C.3   SAMPLING PERIODS

The experimental study was conducted over four weeks. During this period, fine $(0.01–2.5 \ \mu m \ [3.93701 \times 10^{-7}–9.84252 \times 10^{-5}$ in.]) particle mass and concentration and particle chemistry (sulfate, nitrate, and elemental carbon) were measured for different HVAC filter ratings (MERV 7, 11, and 14). The measurements were separate

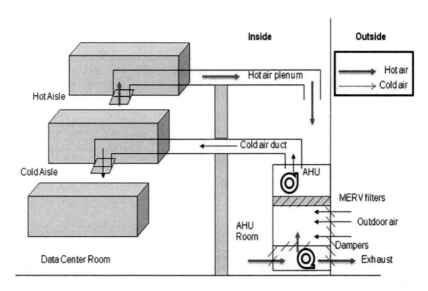

Figure C.1   Layout of the data center and AHU (chiller system not in illustration).

for the hours when the air-side economizer was used (economizer-on mode) and not used (economizer-off mode). A single aerosol measurement setup switched between indoor and outdoor air. Chemical species were measured using two separate filter setups to sample outdoor and indoor air. Each setup had two flow lines: one line sampled during the economizer-on mode and the other during the economizer-off mode. Each setup had a batch of filters to determine chemical speciation of the particulates. Temperature, relative humidity (RH), fan energy, and HVAC energy were also recorded throughout the study. Corrosion classification coupons were placed inside the data center throughout the 30-day period.

C.4   EXPERIMENTAL METHODS

A single set of instruments measured indoor and outdoor fine (0.01–2.5 µm [3.93701 × 10^{-7}–9.84252 × 10^{-5} in.]) particle concentration. The instruments consisted of two separate sample lines one for indoor air and the other for outdoor air. They connected to a valve that switched between indoor and outdoor air. The inlet airstream led to a 2.5 µm [9.84252 × 10^{-5} in.] cyclone precut that fed to an optical particle counter and a dual-wavelength aethalometer.

For measuring PM2.5 particle chemistry, there were two separate setups: one for outdoor air and the other for indoor air. Figure C.2 shows the single-filter setup. Each setup consisted of a cyclone precut (2.5 µm [9.84252 × 10^{-5} in.] with a flow rate of 25 L/min [6.604 gal/min]), which divided into two flow lines: one line sampled during economizer-on mode and the other during the economizer-off mode. A flow line had two filter holders in parallel. One filter holder held the dual quartz filters for total carbon and black carbon concentrations. The second filter holder consisted of a honeycomb denuder followed by teflon, cellulose, and nylon filters. The teflon filter measured total mass, cellulose filter gaseous nitric acid, and nylon gaseous ammonia. All of the filters were equilibrated and weighed before and after being used in the filter setup.

Temperature and RH were measured in the aisles as well as outside the data center. The variations in fan and HVAC energy due to usage of different MERV filters were recorded in the EMS of the data center. Corrosion classification coupons were placed inside and outside the data center to measure the corrosion from particulates and gases.

C.5   ANALYSIS AND RESULTS

The data center IAQ study was recently completed. The analysis is presently a work in progress and will be completed shortly with the results published soon thereafter. A copy of the report will be made available at http://hightech.lbl.gov/datacenters.html and www.osti.gov/bridge/.

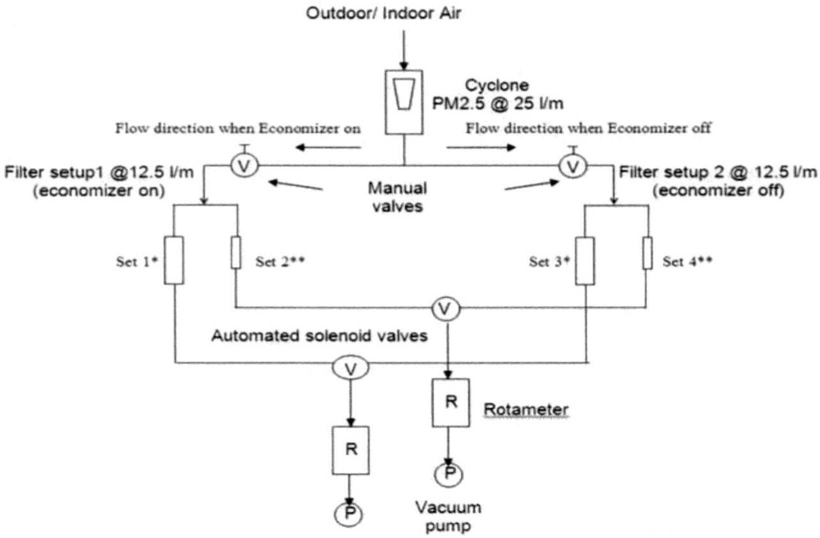

(I-P units: 25 L/m [6.6 gal/min] and 12.5 L/m [3.3 gal/min])

    Set 1: Teflon filter (total mass), followed by nylon filter (nitric acid), and citric-acid-impregnated cellulose filter (ammonia).
    Set 2: Two quartz filters.
    Set 3: Teflon filter (total mass), followed by nylon filter (nitric acid), and citric-acid-impregnated cellulose filter (ammonia).
    Set 4: Two quartz filters.

Figure C.2   Filter setup for indoor and outdoor particulate matter (PM).

References and Bibliography

REFERENCES

ASHRAE. 2006. *Liquid Cooling Guidelines for Datacom Equipment Centers.* Atlanta: American Society of Heating, Refrigerating and Air-Conditioning Engineers, Inc.

ASHRAE. 2007a. *ANSI/ASHRAE Standard 52.2-2007, Method of Testing General Ventilation Air-Cleaning Devices for Removal Efficiency by Particle Size.* Atlanta: American Society of Heating, Refrigerating and Air-Conditioning Engineers, Inc.

ASHRAE. 2007b. *ANSI/ASHRAE Standard 127-2007, Method of Testing for Rating Computer and Data Processing Room Unitary Air Conditioners.* Atlanta: American Society of Heating, Refrigerating and Air-Conditioning Engineers, Inc.

ASHRAE. 2007c. *ANSI/ASHRAE Standard 62.1-2007, Ventilation for Acceptable Indoor Air Quality.* Atlanta: American Society of Heating, Refrigerating and Air-Conditioning Engineers, Inc.

ASHRAE. 2007d. *ANSI/ASHRAE/IESNA Standard 90.1-2007, Energy Standard for Buildings Except Low-Rise Residential Buildings.* Atlanta: American Society of Heating, Refrigerating and Air-Conditioning Engineers, Inc.

ASHRAE. 2008. *2008 ASHRAE Handbook—HVAC Systems and Equipment,* Chapter 21. Atlanta: American Society of Heating, Refrigerating and Air-Conditioning Engineers, Inc.

ASHRAE. 2009a. *Design Considerations for Datacom Equipment Centers,* 2d ed. Atlanta: American Society of Heating, Refrigerating and Air-Conditioning Engineers, Inc.

ASHRAE. 2009b. *Thermal Guidelines for Data Processing Environments,* 2d ed. Atlanta: American Society of Heating, Refrigerating and Air-Conditioning Engineers, Inc.

ASHRAE. 2009c. *Best Practices for Datacom Facility Energy Efficiency*, 2d. ed. Atlanta: American Society of Heating, Refrigerating and Air-Conditioning Engineers, Inc.

Bailey, M., M. Eastwood, T. Grieser, L. Borovick, V. Turner, and R.C. Gray. 2007. *Special Study: Data Center of the Future*, IDC #06C4799. Framingham, MA: International Data Group.

Bell, A. 2000. *HVAC Equations, Data, and Rules of Thumb*. New York: McGraw-Hill.

Bellcore. 1995. *GR-63-CORE, Network Equipment-Building System (NEBS) Requirements: Physical Protection*. Issue 1, October. New Jersey: Bellcore.

Data Center Journal. 2008. Have you heard of ISA71? If you have not you should. http://datacenterjournal.com/index.php?option=com_content&task=view&id=2013&Itemid=43.

European Parliament. 2003. Directive 2002/95/EC of the European Parliament and of the Council of 27 January 2003 on the Restriction of the use of Certain Hazardous Substances on Electrical and Electronic Equipment. Official Journal L 037 (February 13, 2003):19–23.

Ganguly, S. 2009. High performance buildings: data center indoor air quality using an air-side economizer. Lawrence Berkeley National Laboratory, Berkeley, CA. http://hightech.lbl.gov/datacenters.html.

Hanley, J.T., D.S. Ensor, D.D. Smith, and L.E. Sparks. 1994. Fractional aerosol filtration efficiency of in-duct ventilation air cleaners. *Indoor Air* 4(3):169–78.

Herrlin, M. 1997. The pressurized telecommunications central office: IAQ and energy consumption. Healthy Buildings/IAQ'97, September 27–October 2, Washington, DC.

Herrlin, M. 2007. Telephone communication with the authors, June.

IEC. 2002. *IEC 60721-3-3, Classification of Groups of Environmental Parameters and Their Severities—Stationary Use at Weather-Protected Locations*. Geneva: International Electrotechnical Commission.

Institute of Environmental Sciences. 1992. *U.S. Federal Standard 209E, Airborne Particulate Cleanliness Classes in Cleanrooms and Clean Zones*. Illinois: Institute of Environmental Sciences.

ISA. 1985. *ANSI/ISA S71.04-1985, Environmental Conditions for Process Measurement and Control Systems: Airborne Contaminants*. Research Triangle Park, NC: International Society of Automation.

ISO. 1999. *ISO 14644-1, Cleanrooms and Associated Controlled Environments—Part 1: Classification of Air Cleanliness*. Geneva: International Organization for Standardization.

ISO. 2000. *ISO 14644-2, Cleanrooms and Associated Controlled Environments—Part 2: Specifications for Testing And Monitoring To Prove Continued Compliance with ISO 14644-1*. Geneva: International Organization for Standardization.

Johansson, L. 1985. Laboratory study of the influence of NO_2 and combination of NO_2 and SO_2 on the atmospheric corrosion of different metals. *Electrochem. Soc Extended Abstracts* 85(2):221–22.

Keady, P.B. 2000. Getting Data You Need With Particle Measurements. *Indoor Environment Connections* 2(1).

Krzyzanowski, M.E., and B.T. Reagor. 1991. Measurement of potential contaminants in data processing environments. *ASHRAE Transactions* 97(1):464–76.

Lighthouse. 2006. How to select a particle counter for my cleanroom. *Lighthouse Application Note.* Lighthouse Worldwide Solutions. www.golighthouse.com/ApplicationNotes/LetterSize How_to_Select_a_Particle_Counter_for_my_Cleanroom.pdf.

Litvak, A., A.J. Gadgil, and W.J. Fisk. 2000. Hygroscopic fine mode particle deposition on electronic circuits and resulting degradation of circuit performance: An experimental study. *Indoor Air* 10(1):47–56.

McMurry, P.H., M.F. Shepherd, and J.S. Vickery, eds. 2004. *Particulate Matter Science for Policy Makers: A NARSTO Assessment.* Cambridge, England: Cambridge University Press.

Montoya, J. 2002. Effect of Dew Point and Relative Humidity in Electrostatic Charge Control. Sematech Electrostatic Discharge Impact and Control Workshop, October 14, Austin, TX.

NFPA. 2006. *NFPA 750: Standard on Water Mist Fire Protection Systems.* Quincy, MA: National Fire Protection Agency.

NFPA. 2007. *NFPA 13: Standard for the Installation of Sprinkler Systems.* Quincy, MA: National Fire Protection Agency.

NFPA. 2008. *NFPA 2001: Standard on Clean Agent Fire Extinguishing Systems.* Quincy, MA: National Fire Protection Agency.

NFPA. 2009. *NFPA 12A: Standard on Halon 1301 Fire Extinguishing Systems.* Quincy, MA: National Fire Protection Agency.

Ortiz, S. 2006. Data center cleaning services. *Processor. Soc.* 28(14):4.

Riley, W.J., T.E. McKone, A.C.K. Lai, and W.W. Nazaroff. 2002. Indoor particulate matter of outdoor origin: Importance of size-dependent removal mechanisms. *Environmental Science and Technology* 36(2):200–207.

Schueller, R. 2007. Creep corrosion on lead-free printed circuit boards in high sulfur environments. Surface Mount Technology Association, October, Orlando, FL.

Seinfeld, J.H., and S.N. Pandis. 1998. *Atmospheric Chemistry and Physics.* New York: Wiley.

Shehabi, A., A. Horvath, W. Tschudi, A.J. Gadgil, and W.W. Nazaroff. 2008. Particle concentrations in data centers. *Atmospheric Environment* 42:5978–90.

Stack, F. 2010. Untitled forthcoming paper. 2010 ASHRAE Winter Conference, Orlando, FL, January 23–27.

Telcordia. 2006. GR-63-CORE, Network equipment-building systems requirements: Physical protection. *Telcordia Technologies Generic Requirements, Issue 3*. Piscataway, NJ: Telcordia Technologies, Inc.

Volpe, L., and P.J. Peterson. 1989. Atmospheric sulfdization of silver in tubular corrosion reactor. *Corrosion Science* 29(10):1179–96.

BIBLIOGRAPHY

DiNenno, P. 1998. Direct halon replacement agents and systems. In NFPA's *Fire Protection Handbook, Eighteenth Edition*, pp. 6-318–20. Quincy, MA: National Fire Protection Agency.

EPA. 2007. Air Data: Access to air pollution data. Environmental Protection Agency, Washington, DC. www.epa.gov/air/data.

FEMA. 2009. Mapping Information Platform. Federal Emergency Management Agency, Washington, DC. https://hazards.fema.gov/femaportal/wps/portal.

Great Lakes. 1997a. *Understanding the Thermal Decomposition of FM-200 and the Effect on People and Equipment*. Middlebury, CT: Great Lakes Chemical Corp. (now Chemtura Corporation).

Great Lakes. 1997b. *Understanding Current Fire Protection Standards and FM-200 Performance*. Middlebury, CT: Great Lakes Chemical Corp. (now Chemtura Corporation).

Great Lakes. 1998. *Special Hazards Testing by the Loss Prevention Council*. Middlebury, CT: Great Lakes Chemical Corp. (now Chemtura Corporation).

Great Lakes. 2003. *FM-200 and Cultural Heritage Materials: A Pilot Study in Exposure and Discharge Effects*. Middlebury, CT: Great Lakes Chemical Corp. (now Chemtura Corporation).

Hughes Associates. 1995. *Hazard Assessment of Thermal Decomposition Products of FM-200 in Electronics and Data Processing Facilities*. Baltimore: Hughes Associates, Inc.

Kucera V., and E. Mattsson. 1987. Atmospheric Corrosion. *Corrosion Mechanisms*, ed. F. Mansfel, 214. New York: Marcel Dekker.

LPC. 1996. *Halon Alternatives: A Report on the Fire Extinguishing Performance Characteristics of Some Gaseous Alternatives to Halon 1301*. London: Loss Prevention Council.

Muller, C.O. 1999. Control of corrosive gases to avoid electrical equipment failure. PITA Annual Conference.

NASA. 2009. Tin whisker (and other metal whisker) homepage. National Aeronautic and Space Administration, Washington, DC. http://nepp.nasa.gov/whisker.

NFPA. 2008. *NFPA 2001, Standard on Clean Agent Fire Extinguishing Systems*. Quincy, MA: National Fire Protection Agency.

NFPA. 2009a. *NFPA 76, Standard for the Fire Protection of Telecommunication Facilities*. Quincy, MA: National Fire Protection Agency.

NFPA. 2009b. *NFPA 75, Standard for the Fire Protection of Information Technology Equipment*. Quincy, MA: National Fire Protection Agency.

NFPA. 2009c. *NFPA 12A, Standard on Halon 1301 Fire Extinguishing Systems*. Quincy, MA: National Fire Protection Agency.

Rice, D.W., P.B.P Phipps, and R. Tremoureux. 1980. Atmospheric corrosion of nickel. *Journal of Electrochem. Soc.* 127(3):563–68.

Rice, D.W., P. Peterson, E.B. Rigby, P.B.P. Phipps, R.J. Cappell, and R. Tremoureux. 1981. Atmospheric corrosion of copper and silver. *Journal of Electrochem. Soc.* 128(2):275–84.

Shields, H.C., and C.J. Weschler. 1998. Are indoor air pollutants threatening the reliability of your electronic equipment? *Heating/Piping/Air Conditioning Engineering* 70(5):46–54.

Su, J., A. Kim, and J. Mawhinney. 1996. Review of total flooding gaseous agents as Halon 1301 substitutes. *Journal of Fire Protection Engineering* 8(2):45–64.

Weschler, C.J. 1991. Predictions of benefits and costs derived from improving indoor air quality in telephone switching offices. *Indoor Air* 1(1):65–78.

Glossary of Terms

accumulation mode particles: also known as *secondary particles*, which are formed in the atmosphere owing to both the chemical and physical processes that take place with the interactions of primary gaseous emissions. The primary gaseous emissions are injected into the atmosphere by combustion processes such as from a car or from a coal burning plant.

advection: the transfer of heat by the horizontal movement of air.

aethalometer: an instrument used to measure real-time particulate matter concentration.

anthropogenically: creation of particulate matter from humans.

arrestance: the amount of synthetic dust a filter is able to capture.

coulometric reduction: a method used to measure the copper corrosion rate on metal surfaces.

CRAC: computer room air conditioner; generally refers to computer room cooling units that utilize dedicated compressors and refrigerant cooling coils rather than chilled-water coils.

CRAH: computer room air handler; generally refers to computer room cooling units that utilize chilled-water coils for cooling rather than dedicated compressors.

CSU: channel service unit; a terminator and error correction device that is used to connect a router to a digital carrier. It is used in conjunction with a data service unit to act as a network interface card.

cyclone precut: a centrifugal particle collector that uses cyclonic action to remove particles from an airstream. The size of the particles removed depends on the flow rate of the incoming airstream.

DASD: direct access storage device; also known as a *hard disk drive.*

data center: any datacom environment dedicated to housing and operating equipment used in the transfer, storage, and processing of electrical signals for communication or computation.

    NOTE: A data center is a datacom environment, but a datacom environment is not necessarily a data center. The operative differentiator is *dedicated environment.*

datacom: a term that is used as an abbreviation for the data and communications industry.

datacom environment: any area used to house or mount equipment used in the transfer, storage, and processing of electrical signals for communication or computation.

datacom equipment center: a building or portion of a building where the primary function is to house a computer room and its support areas. Datacom equipment centers typically contain high-end servers and communication and storage products with mission-critical functions.

datacom technology: any equipment used in the transfer, storage, and processing of electrical signals for communication or computation. Equipment may or may not actively alter the electrical signals (e.g., punch down block and cabling versus server or network switch).

deliquesce: salts that become a liquid by absorbing moisture from the air.

dew point: the temperature at which water vapor has reached the saturation point (100% relative humidity).

DSU: data service unit; a DSU is a digital packet converter that is used to ensure proper data formatting between a router and a digital carrier. It is used in conjunction with a channel service unit to act as a network interface card.

dust spot efficiency: a measure of the ability of the filter to remove atmospheric dust from the test air.

DX: direct expansion; a system is one where the cooling effect is obtained directly from the refrigerant. It typically incorporates a compressor and in most cases the refrigerant undergoes a change of state in the system.

EDX: energy-dispersive X-ray; a spectroscopy technique that is an analytical technique used for the elemental analysis or chemical characterization of a particulate matter sample.

EMS: energy management system; a collection of integrated hardware and software tools that allow a datacom equipment center operator to monitor, control, and optimize the environment.

EPO: emergency power off; a single switch or device used to quickly remove electricity or power from IT and infrastructure equipment in the datacom environment. The use of an EPO may be dictated by fire or electrical codes under the auspices of emergency shutdown or disconnecting means.

GC/MS: gas chromatography/mass spectrometry; a method that combines aspects of gas-liquid chromatography and mass spectrometry to positively identify minute concentrations of elements within a test sample.

HEPA: high-efficiency particulate air.

HEPA filter: filters designed to remove at least 99.97% or more of all airborne particles 0.3 micrometers (μm) (1.18×10^{-5} in.) or larger from the air that passes through the filter. There are different levels of cleanliness, and some HEPA filters are designed for even higher removal efficiencies and/or removal of smaller particles.

HVAC: heating, ventilation, and air conditioning; in the datacom equipment center, HVAC systems control the ambient environment (i.e., temperature, humidity, airflow, and air filtering) and must be planned for and operated along with other datacom center equipment such as computing hardware, cabling, data storage, fire protection, physical security systems, and power.

hygroscopic: substances that can attract, absorb, and retain moisture from the atmosphere.

IEC: International Electrotechnical Commission; a not-for-profit, nongovernmental international standards organization that prepares and publishes international standards for all electrical, electronic, and related technologies—collectively known as *electrotechnology*. IEC standards cover a vast range of technologies from power generation, transmission, and distribution to home appliances and office equipment, semiconductors, fiber optics, batteries, solar energy, nanotechnology, and marine energy, as well as many others. The IEC also manages three global conformity assessment systems that certify whether equipment, systems, or components conform to international standards.

infiltration: flow of outdoor air into a building through cracks and other unintentional openings and through the normal use of exterior doors for entrance and egress; also know as *air leakage into a building*.

ISO: International Organization for Standardization; an international standard-setting body composed of representatives from various national standards organizations. The organization promulgates worldwide proprietary industrial and commercial standards. While ISO defines itself as a nongovernmental organization, its ability to set standards that often become law, either through treaties or national standards, makes it more powerful than most nongovernmental organizations. In practice, ISO acts as a consortium with strong links to governments.

MERV: minimum efficiency reporting value; the MERV rating on an air filter describes its efficiency as a means of reducing the level of 3 to 10 micron-sized particles in air that passes through the filter. Higher MERV means higher filter efficiency. The purpose of the MERV standard is to permit an equal comparison of the filtering efficiency of various air filters.

QCM: quartz crystal microbalance; a precision instrument used to measure very small masses by a change in frequency of a quartz crystal resonator.

PM: particulate matter; a generic term used to describe a complex group of air pollutants that vary in size and composition, depending upon the location and time of its source. The PM mixture of fine airborne solid particles and liquid droplets (aerosols) include components of nitrates, sulfates, elemental carbon, organic carbon compounds, acid aerosols, trace metals, and geological material. Some of the aerosols are formed in the atmosphere from gaseous combustion by-products such as volatile organic compounds, oxides of sulfur (SO_x), and nitrogen oxides (NO_x). The size of PM can vary from coarse wind-blown dust particles to fine particles directly emitted or formed from chemical reactions occurring in the atmosphere.

RH: relative humidity; a ratio of the partial pressure or density of water vapor to the saturation pressure or density, respectively, at the same dry-bulb temperature and barometric pressure of the ambient air. At 100% RH, the dry-bulb, wet-bulb, and dew-point temperatures are equal.

RoHS: restriction of hazardous substances; RoHS regulations are European Union regulations enforceable after July 1, 2006, that set maximum concentration limits on hazardous materials used in electrical and electronic equipment. The substances are lead, mercury, cadmium, hexavalent chromium, polybrominated biphenyls, and polybrominated diphenyl ethers flame retardants. There are some datacom equipment exemptions, depending on the device being manufactured. There are some exceptions such as lead in solders used in high-reliability applications for which there is no known substitute. Mercury is permitted in limited quantities in some fluorescent lamps yet unrestricted in other types.

SAN: storage area network; the SAN architecture allows remote storage to appear as if it is locally attached to a server.

sedimentation: deposition or settling of particles via gravity, centrifugal force, or electric field.

SEM: scanning electron microscope; a sophisticated microscope that uses a high-energy beam. It collects signals off of a surface, which indicate material properties such as composition and electrical conductivity.

telecom: abbreviation for telecommunications.

TSP: total suspended particulate; a measured parameter of the solid particles (e.g., wood, process dust, and smoke) found in air emissions. These tiny airborne particles or aerosols that are less than 100 μm (3.94×10^{-3} in.) are collectively referred to as *total suspended particulate matter*. These particles constantly enter the atmosphere from many sources. For example, they result from motor vehicle use, combustion products from space heating, industrial processes, power generation, soil, bacteria and viruses, fungi, molds and yeast, pollen, salt particles from evaporating sea water, and many others.

VCT: vinyl composition tile; VCT is a finished flooring material used primarily in commercial and institutional applications. Vinyl tiles are composed of colored vinyl chips formed into solid sheets of varying thicknesses (1/8 in. [3.2 mm] is most common) by heat and pressure and cut into 12 in. (305 mm) squares. Tiles are applied to a smooth, leveled sub-floor using a specially formulated vinyl adhesive that remains tacky but does not dry completely. Once in place, tiles are typically waxed and buffed using materials and equipment specially formulated for this application.

VOCs: volatile organic compounds; organic chemical compounds that have high enough vapor pressures under normal conditions to significantly vaporize and enter the atmosphere. A wide range of carbon-based molecules, such as aldehydes, ketones, and other light hydrocarbons, are VOCs.

whiskers: crystalline metalurigical phenomenon whereby iron, tin, and zinc grow tiny hairs that can become airborne under certain conditions and settle in datacom equipment.

Index